Material Characterization of Field-Cast Connection Grouts

January 2013

NTIS Accession No. PB2013-130231

FHWA Publication No. FHWA-HRT-13-041

U.S. Department of Transportation
Federal Highway Administration

FOREWORD

The maintenance and reconstruction needs facing our highway transportation system are significant. There is a strong need to construct robust, durable bridge structures without unduly restricting the capacity of the existing highway network. Accelerated bridge construction activities are gaining favor around the country, most notably in jurisdictions where the new construction concepts afford reduced user impacts concurrent with high quality finished products. The Federal Highway Administration's emphasis on the use of Prefabricated Bridge Elements and Systems (PBES) as a means to facilitate accelerated bridge construction has focused national attention through the Every Day Counts initiative on a set of emerging concepts. PBES, wherein large portions of the bridge are fabricated off-site then are transported and assembled rapidly at the bridge site, is a technology that offers significant benefits in terms of component quality and construction site safety; however, PBES frequently relies heavily on the use of field-cast grouts to complete connections between components. Recent advancements in the rheological, mechanical, and durability behaviors of field-cast grout-type materials has resulted in owners facing a wide range of options when considering the appropriate grout for a particular application. This report presents the results of a wide-ranging investigation into the material characteristics of eight prebagged grouts, providing the basis for a broader understanding of the short- and long-term properties that an owner could anticipate experiencing with these materials.

This report corresponds to the TechBrief titled "Material Characterization of Field-Cast Connection Grouts" (FHWA-HRT-13-042). This report is being distributed through the National Technical Information Service for informational purposes. The content in this report is being distributed "as is" and may contain editorial or grammatical errors.

Notice

This document is disseminated under the sponsorship of the U.S. Department of Transportation in the interest of information exchange. The U.S. Government assumes no liability for the use of the information contained in this document.

The U.S. Government does not endorse products or manufacturers. Trademarks or manufacturers' names appear in this report only because they are considered essential to the objective of the document.

Quality Assurance Statement

The Federal Highway Administration (FHWA) provides high-quality information to serve Government, industry, and the public in a manner that promotes public understanding. Standards and policies are used to ensure and maximize the quality, objectivity, utility, and integrity of its information. FHWA periodically reviews quality issues and adjusts its programs and processes to ensure continuous quality improvement.

TECHNICAL REPORT DOCUMENTATION PAGE

1. Report No. FHWA-HRT-13-041	2. Government Accession No. NTIS PB2013-130231	3. Recipient's Catalog No.
4. Title and Subtitle Material Characterization of Field-Cast Connection Grouts		5. Report Date January 2013
		6. Performing Organization Code:
7. Author(s) Matthew K. Swenty and Benjamin A. Graybeal		8. Performing Organization Report No.
9. Performing Organization Name and Address Office of Infrastructure Research & Development Federal Highway Administration 6300 Georgetown Pike McLean, VA 22101-2296		10. Work Unit No.
		11. Contract or Grant No.
12. Sponsoring Agency Name and Address Office of Infrastructure Research & Development Federal Highway Administration 6300 Georgetown Pike McLean, VA 22101-2296		13. Type of Report and Period Covered Final Report: 2010-2012
		14. Sponsoring Agency Code HRDI-40

15. Supplementary Notes
The research discussed herein was completed at the Turner-Fairbank Highway Research Center. Portions of the work were completed by PSI, Inc. under contract DTFH61-10-D-00017. Other parts of the testing discussed herein were completed by Global Consulting, Inc. under contract DTFH61-07-C-00011. Matthew Swenty, formerly employed by PSI, Inc. and currently employed by the Virginia Military Institute, was the co-Principal Investigator on this project with Ben Graybeal who leads the FHWA Structural Concrete Research Program.

16. Abstract
Accelerated bridge construction methods can help increase safety and minimize the inconveniences to the traveling public. Many new construction methods have been investigated and implemented using prefabricated subassemblies on bridges. These methods have shown great promise because prefabricated components can be produced in a controlled environment, resulting in superior products that both allow for expedited construction schedules and display robust, durable performance. The most critical field construction process for prefabricated subassemblies is the completion of the connections. Short-term construction-related issues along with long-term serviceability problems have been attributed to a variety of causes, including construction techniques, materials, and poor designs. This research effort investigated the performance of a variety of different material categories that may be used in field-cast highway infrastructure connections. The objective of this research is to evaluate grout-type materials for potential use in field-cast connections deployed as part of the prefabricated bridge elements and systems (PBES) bridge construction concept. Topics investigated included mechanical strength, shrinkage, bond strength, durability, and cost. The results demonstrate that a wide range of grout performances are achievable, and thus owners should carefully consider the parameters critical to the successful completion of any pending project.

This report corresponds to the TechBrief titled "Material Characterization of Field-Cast Connection Grouts" (FHWA-HRT-13-042).

17. Key Words Field-Cast Grout; Prefabricated Bridge Elements and Systems (PBES); Mechanical Strength; Shrinkage; Bond Strength; Durability		18. Distribution Statement No restrictions. This document is available through the National Technical Information Service, Springfield, VA 22161.		
19. Security Classif. (of this report) Unclassified	20. Security Classif. (of this page) Unclassified		21. No. of Pages 91	22. Price N/A

Form DOT F 1700.7 (8-72)   Reproduction of completed page authorized

SI* (MODERN METRIC) CONVERSION FACTORS

APPROXIMATE CONVERSIONS TO SI UNITS

Symbol	When You Know	Multiply By	To Find	Symbol
LENGTH				
in	inches	25.4	millimeters	mm
ft	feet	0.305	meters	m
yd	yards	0.914	meters	m
mi	miles	1.61	kilometers	km
AREA				
in^2	square inches	645.2	square millimeters	mm^2
ft^2	square feet	0.093	square meters	m^2
yd^2	square yard	0.836	square meters	m^2
ac	acres	0.405	hectares	ha
mi^2	square miles	2.59	square kilometers	km^2
VOLUME				
fl oz	fluid ounces	29.57	milliliters	mL
gal	gallons	3.785	liters	L
ft^3	cubic feet	0.028	cubic meters	m^3
yd^3	cubic yards	0.765	cubic meters	m^3

NOTE: volumes greater than 1000 L shall be shown in m^3

Symbol	When You Know	Multiply By	To Find	Symbol
MASS				
oz	ounces	28.35	grams	g
lb	pounds	0.454	kilograms	kg
T	short tons (2000 lb)	0.907	megagrams (or "metric ton")	Mg (or "t")
TEMPERATURE (exact degrees)				
°F	Fahrenheit	5 (F-32)/9 or (F-32)/1.8	Celsius	°C
ILLUMINATION				
fc	foot-candles	10.76	lux	lx
fl	foot-Lamberts	3.426	candela/m^2	cd/m^2
FORCE and PRESSURE or STRESS				
lbf	poundforce	4.45	newtons	N
lbf/in^2	poundforce per square inch	6.89	kilopascals	kPa

APPROXIMATE CONVERSIONS FROM SI UNITS

Symbol	When You Know	Multiply By	To Find	Symbol
LENGTH				
mm	millimeters	0.039	inches	in
m	meters	3.28	feet	ft
m	meters	1.09	yards	yd
km	kilometers	0.621	miles	mi
AREA				
mm^2	square millimeters	0.0016	square inches	in^2
m^2	square meters	10.764	square feet	ft^2
m^2	square meters	1.195	square yards	yd^2
ha	hectares	2.47	acres	ac
km^2	square kilometers	0.386	square miles	mi^2
VOLUME				
mL	milliliters	0.034	fluid ounces	fl oz
L	liters	0.264	gallons	gal
m^3	cubic meters	35.314	cubic feet	ft^3
m^3	cubic meters	1.307	cubic yards	yd^3
MASS				
g	grams	0.035	ounces	oz
kg	kilograms	2.202	pounds	lb
Mg (or "t")	megagrams (or "metric ton")	1.103	short tons (2000 lb)	T
TEMPERATURE (exact degrees)				
°C	Celsius	1.8C+32	Fahrenheit	°F
ILLUMINATION				
lx	lux	0.0929	foot-candles	fc
cd/m^2	candela/m^2	0.2919	foot-Lamberts	fl
FORCE and PRESSURE or STRESS				
N	newtons	0.225	poundforce	lbf
kPa	kilopascals	0.145	poundforce per square inch	lbf/in^2

*SI is the symbol for the International System of Units. Appropriate rounding should be made to comply with Section 4 of ASTM E380.
(Revised March 2003)

TABLE OF CONTENTS

CHAPTER 1. INTRODUCTION 1
    INTRODUCTION 1
    OBJECTIVE 1
    SUMMARY OF APPROACH 1
    OUTLINE OF REPORT 1

CHAPTER 2. LITERATURE REVIEW 3
    INTRODUCTION 3
    PAST RESEARCH RESULTS 3
    RESEARCH METHODS 4
    SUMMARY 5

CHAPTER 3. MATERIAL TESTING PROGRAM 7
    RESEARCH PLAN 7
    TESTING MATRIX 7
    BATCHING, CASTING, AND CURING SPECIMENS 9
        G1, G2, G3 and T1 11
        M1 12
        E1 13
        U1 13
        U2 14
        C1 14
    CONSTRUCTABILITY 15
        Workability 15
        Set Time 18
        Cost 20
    MATERIAL PROPERTIES 21
        Unit Weight 21
        Compressive Strength 21
        Split Cylinder Tensile Strength 22
        Modulus of Elasticity 23
        Restrained Shrinkage 24
        Unrestrained Shrinkage 32
    BOND TESTING 35
        Slant Cylinder Compression Test 36
        Splitting Tensile Bond Test 38
        Restrained Shrinkage Bond Test 41
    DURABILITY TESTING 50
        Freeze-Thaw Resistance 50

    Rapid Chloride Penetrability 53
CHAPTER 4. RESULTS **55**
  CONSTRUCTION 55
    Workability 55
    Cleanup 56
    Set Time 56
    Cost 56
  MATERIAL PROPERTIES 57
    Unit Weight 57
    Compressive Strength 58
    Tensile Strength 59
    Modulus of Elasticity 59
    Shrinkage 60
  BOND STRENGTH 61
    Slant Cylinder Compression Test 61
    Splitting Tensile Bond Test 62
  DURABILITY 63
  GRAPHICAL SUMMARY OF GROUT PERFORMANCE RESULTS 64
CHAPTER 5. CONCLUSIONS **65**
  SUMMARY 65
  RECOMMENDATION 65
  ONGOING AND FUTURE RESEARCH 66
  ACKNOWLEDGEMENT 66
REFERENCES **67**
APPENDIX A **71**
  A.1   FIVE STAR GROUT MANUFACTURER'S DATA SHEET 71
  A.2   EMBECO 885 GROUT MANUFATURER'S DATA SHEET 73
  A.3   HARRIS CONSTRUCTION GROUT MANUFACTURER'S DATA SHEET ... 77
  A.4   EUCO CABLE GROUT PTX MANUFACTURER'S DATA SHEET 79
  A.5   SET 45 GROUT MANUFATURER'S DATA SHEET 81
  A.6   FIVE STAR HP EPOXY GROUT MANUFACTURER'S DATA SHEET 85
  A.7   LAFARGE DUCTAL JS1000 MANUFATURER'S DATA SHEET 87
  A.8   LAFARGE DUCTAL JS1100RS MANUFACTURER'S DATA SHEET 89
  A.9   VIRGINIA A4 MIX DESIGN PROPORTIONS 91

LIST OF TABLES

Table 1. Material testing matrix. 7
Table 2. Testing schedule and number of specimens for each test. 9
Table 3. G1 batching information. 11
Table 4. G2 batching information. 11
Table 5. G3 batching information. 12
Table 6. T1 batching information. 12
Table 7. M1 batching information. 12
Table 8. E1 batching information. 13
Table 9. U1 batching information. 13
Table 10. U2 batching information. 14
Table 11. C1 batching information. 15
Table 12. Spread measurements using ASTM C1437-07 methods. 16
Table 13. Mixing notes. 17
Table 14. Penetrometer measurements. 19
Table 15. Bulk material unit cost. 20
Table 16. Unit weights. 21
Table 17. Compressive strength results. 22
Table 18. Splitting tensile strength results. 23
Table 19. Average modulus of elasticity. 23
Table 20. Shrinkage ring results. 26
Table 21. Crack sizes visually observed on shrinkage ring specimens at 28-days. 32
Table 22. Precast concrete properties. 36
Table 23. Slant cylinder bond strength. 38
Table 24. Splitting tensile bond strengths. 40
Table 25. Comparison of splitting cylinder and splitting cylinder bond strengths. 41
Table 26. Restrained shrinkage bond test results. 44
Table 27. Rapid Chloride Ion Penetrability Results. 54

LIST OF FIGURES

Figure 1. Photo. Clockwise from bottom left: The mixer, specimen molds, and specimens curing in the environmental chamber. 10
Figure 2. Photo. Flow measurement of E1. 15
Figure 3. Photo. The paddle mixer and mixing M1. 17
Figure 4. Photo. Cohesion of M1 to steel formwork. 18
Figure 5. Graph. Typical set time development graph. 19
Figure 6. Photo. Typical shrinkage ring. 24
Figure 7. Equation. Net strain in the shrinkage rings. 25
Figure 8. Equation. Net stress in the shrinkage rings. 25
Figure 9. Graph. Strain development in the inner ring over time for G1. 26
Figure 10. Graph. Strain development in the inner ring over time for G2. 27
Figure 11. Graph. Strain development in the inner ring over time for G3. 27
Figure 12. Graph. Strain development in the inner ring over time for M1. 28
Figure 13. Graph. Strain development in the inner ring over time for E1. 29
Figure 14. Graph. Strain development in the inner ring over time for the U1. 30
Figure 15. Graph. Strain development in the inner ring over time for the U2. 30
Figure 16. Graph. Strain development in the inner ring over time for C1. 31
Figure 17. Photo. ASTM C157-08 molds embedded with vibrating wire gages to measure shrinkage. 33
Figure 18. Graph. Unrestrained length change measured via vibrating wire gage. 33
Figure 19. Graph. Unrestrained length change measured via the ASTM C157-08 test method... 35
Figure 20. Photo. Slant cylinder molds and deck concrete halves being sandblasted. 37
Figure 21. Photo. Splitting cylinder bond forms for the first halves; splitting cylinder and slant cylinder forms for the second halves. 39
Figure 22. Photo. Typical 6 in. by 12 in. split cylinder bond specimen and testing setup. 39
Figure 23. Equation. Splitting tensile strength from ASTM C496-04. 40
Figure 24. Photo. Ring setup for the placement of the deck concrete half. 42
Figure 25. Photo. Ring setup for the placement of the second half. 43
Figure 26. Photo. Restrained shrinkage bond test setup. 43
Figure 27. Graph. Restrained shrinkage bond results for G1. 45
Figure 28. Graph. Restrained shrinkage bond results for G2. 45
Figure 29. Graph. Restrained shrinkage bond results for M1. 47
Figure 30. Graph. Restrained shrinkage bond results for E1. 47
Figure 31. Graph. Restrained shrinkage bond results for U1. 48
Figure 32. Graph. Restrained shrinkage bond results for C1. 49
Figure 33. Graph. Relative dynamic modulus of elasticity of freeze/thaw prisms. 51
Figure 34. Graph. Mass change of freeze/thaw prisms. 51
Figure 35. Photo. U2 prism after the completion of 600 freeze/thaw cycles. 52
Figure 36. Photo. G1 prism after the completion of 600 freeze/thaw cycles. 52

Figure 37. Photo. E1 prism after the completion of 600 freeze/thaw cycles. 52
Figure 38. Photo. M1 prism after the completion of 45 freeze/thaw cycles. 53
Figure 39. Graph. Spread measurements using ASTM C1437-07. 55
Figure 40. Graph. Set times based on ASTM C403-08. 56
Figure 41. Graph. Price comparison of the materials. 57
Figure 42. Graph. Unit weights. 57
Figure 43. Graph. Compressive strength results. 58
Figure 44. Graph. Splitting tensile strength results. 59
Figure 45. Graph. Modulus of elasticity results from 28 day tests. 60
Figure 46. Graph. Strain 28 days after casting using the modified ASTM C157 with vibrating wire gages. 61
Figure 47. Graph. Slant cylinder bond strength based on ASTM C882. 62
Figure 48. Graph. Splitting tensile bond strength based on ASTM C496. 63
Figure 49. Graph. Graphical representation of the performance of the tested materials. 64

CHAPTER 1. INTRODUCTION

INTRODUCTION

Accelerated construction methods can help increase construction site safety and minimize the inconveniences to the traveling public. Many new methods have been investigated and implemented using prefabricated subassemblies on bridges. These methods have shown promise because prefabricated components can be produced with great quality control, resulting in superior products that allow for expedited construction schedules. States continue to investigate and advance their respective bridge programs through the use of prefabricated products such as precast bulb tees, full depth precast bridge decks, and box beams.

The most critical field construction process for prefabricated subassemblies is the completion of the connections. Constructability and serviceability problems have arisen in connections on some past projects. These issues have been attributed to a variety of causes, including construction techniques, materials, and poor designs. Much research attention has been placed on making better connections between the components.

One area of investigation relates to the different field-cast materials that might be used to complete the connections. Connections between prefabricated bridge components may exhibit poor performance due to the material selected. This research effort investigated the performance of a variety of different material categories that may be used in non-post-tensioned field-cast highway infrastructure connections.

OBJECTIVE

The objective of this research is to evaluate grout-type materials for potential use in field-cast connections deployed as part of the prefabricated bridge elements and systems (PBES) bridge construction concept.

SUMMARY OF APPROACH

The research consisted of a series of constructability, material characterization, and bond tests among nine unique candidate materials. The materials were chosen as representative samples in the following categories based on published material properties: high strength grouts, deck concretes, magnesium phosphate grout, ultra-high performance concrete, cable grout, and epoxy grout. The testing included many standard tests as published by the American Society for Testing and Materials (ASTM). Other tests were developed based on previous testing experience in combination with standard ASTM test methods. The results were analyzed and used to predict the relative performance of these materials when deployed in this highway bridge construction application.

OUTLINE OF REPORT

This report is divided into five chapters and an appendix. Chapters 1 and 2 provide an introduction and literature review. Chapter 3 presents the results of the material characterization testing program. Chapter 4 presents the analysis of the results and Chapter 5 presents the conclusions of this research program. An appendix is included that contains manufacturer supplied material characterization information.

CHAPTER 2. LITERATURE REVIEW

INTRODUCTION

Researchers continue to focus on advancing the state-of-the-practice for the construction of prefabricated bridge structural elements. Studies have been performed on innovative methods (Issa, Ralph, Thomas, Shaker, & Islam, 2007; Carter, Pilgrim, Hubbard, Poehnelt, & Oliva, 2007; Sullivan, 2007) and alternate designs (Badie, Tadros, & Baishya, 1998). What all the studies have in common is the need to make unique connections between precast concrete components. One of the keys to building a quick and durable bridge superstructure is the connections (Culmo, 2009). The field casting of these connections tends to be the most labor intensive and critical part of making the overall system work successfully. Within these connections there is frequently is a field-cast grout-type material that is used to complete the system. Insufficient performance of the material within the connection can compromise the entire bridge superstructure's performance.

New emphasis has been placed on testing deck level connections for a variety of different precast designs. A sound material is needed to build a variety of connections in order to perform head-to-head tests. As demonstrated below, past research shows that there is not a general consensus on the best type of material to be used in these connections. Furthermore, no prior study has completed a comprehensive assessment of candidate field-cast grout-type materials covering the wide range of relevant materials and characteristics.

PAST RESEARCH RESULTS

The field-cast grout-type materials specified for use in bridge superstructure connections have undergone limited, sporadic research as to their relevance within this application. Most designers specify prepackaged, low shrinkage, high early strength grouts for bridge connections (Culmo, 2009). However, these grouts have not demonstrated consistent performance. When testing various connections in bridge decks researchers have noted that, even under controlled laboratory conditions, shrinkage cracks and durability issues still arise (Markowski, 2005; Swenty, 2009).

Gulyas et al. studied the use of a magnesium phosphate based grout, and a regular cementitious grout for use in shear keys on adjacent box beams in Alaska (Gulyas, Wirthlin, & Champa, 1995). The materials were tested using standard ASTM tests and component tests. The standard tests worked well as a screening process but more representative testing methods were recommended. The magnesium phosphate based grout performed well for the adjacent box beams in Alaska.

Research led by Issa furthered the research by Gulyas et al. by testing four different commercially available materials in component tests. The four materials included two magnesium phosphate based grouts, a standard grout, and a polymer concrete. The focus was on performing shear, tension, and flexure tests on scaled shear keys typically used between adjacent box beams. The results indicated that the magnesium phosphate grouts did not bond well to the substrate concrete, in part because of carbonation effects, and had limited workability. The polymer concrete had the best results and highest compressive strength; however the standard grout performed very well and was easier to use. The authors recommended using the standard grout over the more expensive and harder to use alternatives. It is noted that the recommended material was only one of many similar products marketed as a prebagged grout mix.

Other researchers have developed similar plans for testing materials used in precast component connections. Ma et. al. conducted further investigations into standard grouts and magnesium phosphate grouts for use in component connections. The research focused on finding materials that work well for one and seven day applications (Ma, 2010). The research concluded that the magnesium phosphate grouts can be used successfully. Scholz et. al. studied three cementitious prebagged grouts and magnesium phosphate grout. The researchers studied the bond strength and standard material properties. A ponding test was performed by casting voids in a 4 in. (10.1 cm) deep deck, filling them with grout, and placing a 0.25 in. (8 mm) layer of water on the top surface. Observations were made of any water that flowed through the bonded surface to the bottom of the deck. The grouts with the best bond strength and lowest shrinkage did not predict the best performance in ponding tests. The results of the tests did not lead to a firm recommendation. The grouts all performed differently but magnesium phosphate tended to perform well (Scholz, Wallenfelsz, Ligeron, & Roberts-Wollmann, 2007). A set of guidelines was produced that, when followed, would likely increase the likelihood of good performance.

More recently a group of researchers studied the use of three grouts for use in box girder connections. The grouts were epoxy, cementitious, and cementitious with polypropylene fibers. A series of shear and flexure tests were performed on scaled box girder connections. The study found that epoxy grout bonded better to a base concrete and gained strength faster than a typical grout used on Pennsylvania bridges (DeMurphy, Kim, Sang, & Xiao, 2010).

RESEARCH METHODS
The test methods commonly employed to demonstrate the material-scale performance of these grouts also deserve discussion. Of particular interest are tests on bond strength and shrinkage.

Most prepackaged grout manufacturers use ASTM C827-10 *Change in Height at Early Ages of Cylindrical Specimens of Cementitious Mixtures* to measures height change in the plastic state and report this result as shrinkage (Culmo, 2009). The ASTM C157-08 *Length Change of Hardened Hydraulic-Cement Mortar and Concrete* method of measuring grout shrinkage, which is more commonly used to assess the post-set shrinkage of cementitious composites in the bridge sector, has had limited use. Due to the apparent shrinkage issues encountered in many past bridge applications, the applicability of ASTM C827-10 to this particular grout application deserves further investigation.

The bond strength between field-cast grout-type materials used in bridge connections and the substrate precast concrete has not been systematically investigated. In general, this interface tends to crack first thereby suggesting the bond between the materials is weaker than the tensile strength of each adjoining material (Issa, Yousif, Issa, Kaspar, & Khayyat, 1995).

Past research on bond strength between cementitious materials tends to focus on the strength between a patch material for a bridge deck and the concrete bridge deck. Many different tests have been used including ASTM C882-05 *Bond Strength of Epoxy-Resin Systems Used with Concrete by Slant Shear*, ASTM C496-04 *Splitting Tensile Strength of Cylindrical Concrete*

Specimens, and ASTM C1583-04 *Tensile Strength of Concrete Surfaces and the Bond Strength or Tensile Strength of Concrete Repair and Overlay Materials by Direct Tension (Pull-off Method)*. The results from each test tend to be consistent for its particular application; however the results are not similar among tests and must be interpreted for the particular application (Momayez, Ehsani, Ramezanianpour, & Rajaie, 2005). In general, the bond tests have been used for overlays and thin bonds and not applied to the range of materials used for precast component connections.

One possible way of measuring the bond strength is using a modified splitting tensile strength test based on ASTM C496-04. Previous research has shown that splitting cylinder tests show a very good correlation to tensile strength and have low scatter within the test results. In addition, the tests have been performed on standard 4 in. by 8 in. cylinders and 3 in. by 3 in. by 4 in. prisms with similar results (Geissert, Li, Frantz, & Stephens, 1999).

SUMMARY

A definitive set of guidelines for materials used in precast connections has not been developed. Some research has focused on a particular grout or connection detail location. Other research projects have suggested the use of large component tests to determine whether a grout is appropriate. Many questions remain on the shrinkage and bond strength of the materials used in precast component connections. In addition, some newer materials have not been thoroughly investigated. A comprehensive set of tests are desirable to compare multiple materials to one another and to a baseline, standard bridge deck mix.

CHAPTER 3. MATERIAL TESTING PROGRAM

RESEARCH PLAN

A program was developed to investigate field-cast grout materials that either already are being used or have the potential to be used in modular bridge component connections. The program was designed to characterize nine different unique materials that are readily available as prebagged mixes or standard concrete mixes. The materials included ultra-high performance concrete (UHPC), magnesium phosphate grout, conventional prebagged cementitious grouts, epoxy grout, a bridge deck concrete, and post-tensioning cable grout.

This chapter describes the material testing program. It begins with a description of the materials used in the study and the tests employed. Next the chapter describes the batching, casting, and curing procedures. The results from each individual test are then discussed in the remainder of the chapter.

TESTING MATRIX

The name of each material and its respective manufacturer are listed in Table 1. The program was designed to investigate materials in six different categories: ultra-high performance concrete (U1 and U2), magnesium phosphate grout (M1), conventional prebagged cementitious grouts (G1, G2, and G3), epoxy grout (E1), a bridge deck concrete (C1), and post-tensioning cable grout (T1). Three materials (G1, G2, and G3) were chosen in the conventional cementitious grout category because this category has been investigated in previous studies and is commonly deployed in PBES-type construction projects. It is important to note that, with the possible exception of the U1 and U2 products, other similar commercially available products exist within these categories in the North American market.

Table 1. Material testing matrix.

Material Category	Product Name	Reference Name
Grout	Five Star Grout	G1
Grout	BASF Embeco 885	G2
Grout	Harris Construction Grout	G3
Magnesium Phosphate Grout	BASF Set 45	M1
Epoxy Grout	Five Star HP Epoxy Grout	E1
Cable Grout	Euclid Euco Cable Grout PTX	T1
UHPC	Lafarge Ductal JS1000	U1
UHPC	Lafarge Ductal JS1100RS	U2
Deck Concrete	Virginia A4 Concrete Mix	C1

Material categories were chosen based on past performance, applicability to onsite construction processes, and suitable published properties. G1, G2, and G3 are standard grouts reported to exhibit low shrinkage, good workability, and high early strength. The manufacturers of M1 and E1 report low shrinkage, high early strength, and dimensional stability. These types of grout have also been tested and deployed previously in modular bridge component connections. The manufacturer of T1 reports that it is pumpable, easy to use, and has reasonable strength gain. The ultra-high performance concretes, U1 and U2, are reported to exhibit exceptional mechanical and durability characteristics. These products have gained the attention of many bridge owners as a promising substitute for conventional PBES connection solutions (Graybeal, 2012). C1 is based on a conventional concrete mix design that could be used for a bridge deck and serves as the control within the study.

Each material was cast independent of the others. The objective was to compare the materials based on construction issues, early age properties, long term properties, and bond strength. Some of the desirable properties include early compressive strength gain, high tensile strength, dimensional stability, and strong bond strength. The construction issues included workability, work time, economics, flow, and set time. The bond tests were performed to quantify which materials bond well to a substrate concrete.

Initial tests and data were recorded during the placement of each material. The mix proportions, laboratory environmental conditions, and ease of use were recorded. Observations were also made on the cleanup procedures and price of the materials.

A series of tests, many of them based on standard test methods, were performed on each selected material. All tests were performed under similar conditions in the concrete laboratory at Turner-Fairbank Highway Research Center (TFHRC). The basic material characterization tests were based on ASTM standards for compressive strength (ASTM C39-09a and ASTM C109-02), split cylinder (ASTM C496-04), flow or slump (ASTM C1437-07 and ASTM C143-10), modulus of elasticity (ASTM C469-02), set time (ASTM C403-08), restrained shrinkage (ASTM C1581-09a), and unrestrained shrinkage (ASTM C157-08). A non-standardized method was employed to measure early age unrestrained shrinkage during the first 24 hours after mixing. This was based on the ASTM C157-08 samples with an embedded vibrating wire gage (VWG).

Three tests were used as an indication of the bond strength between each material and a previously cast and cured deck concrete. The first test was based on the standard slant cylinder bond test ASTM C882-05. The second test was based on the split cylinder test ASTM C496-04. The third bond test was based on the restrained shrinkage ring test ASTM C1581-09a.

Two durability tests were also completed on a select set of materials. The standard freeze/thaw resistance test (ASTM C666-03 *Standard Test Method for Resistance of Concrete to Rapid Freezing and Thawing*) and the standard rapid chloride penetrability test (ASTM C1202-10 *Standard Test Method for Electrical Indication of Concrete's Ability to Resist Chloride Ion Penetration*) were completed for four of the grouts.

Data was taken on the schedule shown in Table 2. The first set of tests began immediately after the materials were cast and continued for a minimum of two months. The measurements at 24 hours were used for accelerated construction comparisons while the longer term measurements were used for standard construction schedule comparisons.

Table 2. Testing schedule and number of specimens for each test.

Tests	0 Hr	2 Hr*	6 Hrs*	24 Hrs*	7 Days	14 Days	28 Days
Flow or Slump (ASTM C1437-07 or C143-10)	1						
Set Time (ASTM C403-08)	1 Specimen Minimum – Data until final set						
Compressive Strength (ASTM C39-09a & C109-02)		3	3	3	3		3
Split Cylinder (ASTM C496-04)							3
Modulus of Elasticity (ASTM C469-02)							3
Restrained Shrinkage (ASTM C1581-09a)	1 Specimen – Data every 5 minutes for 56 days						
Unrestrained Shrinkage (ASTM C157-08)	Approximately every 3 days for 100 days						
Early Age Shrinkage (VWG & ASTM C157-08)	colspan: 2 Specimens – Data every 5 minutes for 56 days						
Slant Cylinder (ASTM C882-05)							3
Split Cylinder Bond (Based on ASTM C496-04)							3
Restrained Shrinkage Bond (Based on ASTM C1581-09a)	1 Specimen – Data every 5 minutes for 56 days						
Freeze-Thaw[†] (ASTM C666)	colspan: 3 Specimens for up to 600 Cycles						
Rapid Chloride Penetrability[†] (ASTM C1202)	Results at 57, 126, and 240 days						

* Where applicable - Some materials had not set.
† Tests only completed on grouts G1, E1, M1, and U2

BATCHING, CASTING, AND CURING SPECIMENS

The specimens were all produced and stored at the Turner-Fairbank Highway Research Center concrete laboratory. The laboratory mixing conditions, curing conditions, molds, and testing protocols were kept the same among samples unless noted otherwise. See Figure 1 for photographs of the molds, mixer, and specimen storage location.

Deviations in mixing occurred because of mixer capacity and work time. Six of the nine materials were mixed in a single batch under the manufacturer's recommendations for a fluid mix suitable for pouring tight joints. U1 and U2 required two batches due to volume and mixing limitations of the mixer. M1 required a large number of mixes because of the short workability time and number of specimens required. All the materials were immediately placed in molds. Aside from M1, a pan mixer was used to mix the materials. M1 was mixed with a paddle mixer inside plastic buckets.

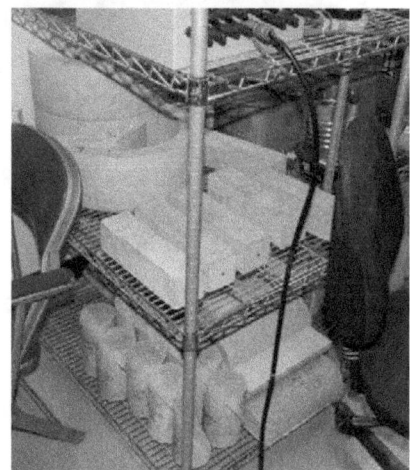

Figure 1. Photo. Clockwise from bottom left: The mixer, specimen molds, and specimens curing in the environmental chamber.

Casting location and curing conditions were constant unless there was a specific deviation in the manufacturer's recommendations. The unrestrained shrinkage bars and all of the shrinkage rings were cast and cured in a temperature and humidity controlled curing room. This was done because the data acquisition setup could not easily be moved between the concrete mixing room and the curing room. The remaining specimens were cast inside the concrete laboratory mixing room, held for 24 hours in that room, demolded, and then immediately placed in the curing room. The specimens were all covered in moist burlap and plastic for the first 24 hours regardless of their curing location. M1 was the exception as it did not require a moist burlap layer according to the manufacturer. Once all specimens were inside the curing room, the coverings were removed and the specimens were left to cure in the controlled environment. All specimens remained in the curing room until the end of the tests. The curing room was held at a humidity of 45% ± 5% and a temperature of 74°F ± 4°F (23°C ± 2°C).

The curing for the freeze-thaw prisms and the rapid chloride penetrability cylinders deviated from the curing described above. These prisms and cylinders were placed in a lime water bath after demolding. In accordance with the ASTM C666 test method, the prisms were removed from the bath after 14 days so that the freeze-thaw testing could begin. The cylinders remained in the bath until each set of ASTM C1202 testing was ready to commence.

G1, G2, G3 and T1

The batching information for G1, G2, G3, and T1 are in Table 3 through Table 6, respectively. The guidelines given by the manufacturers were followed as shown in Appendix A.1 through Appendix A.4. The temperatures at casting were all close to 70°F (21.1°C), the proportions were for fluid mixes, and the mix times fell within the published ranges.

Each material was mixed according to the manufacturers' recommendations. They were initially mixed for 5 minutes and then inspected for proper consistency. G2 was found to have some clumps of unhydrated material that ranged in size from approximately 1/2 in. to 2 in. (1.2 cm to 5.1 cm). The balls were physically broken by hand and then the material was mixed an additional 1 minute. T1 was not fluid after 5 minutes, therefore mixing continued. There was no manufacturer recommendation for maximum mix time; therefore, T1 was mixed for 12.5 minutes, at which point when the mix behaved fluidly. It should be noted that no water was added to any of the mixes beyond the maximum limits for fluid mixes.

Table 3. G1 batching information.

Placement Date	Grout, lbs (kg)	Water, lbs (kg)	Mix Time, Minutes	Lab Temp., °F (°C)	Grout Temp. After Mixing, °F (°C)
16Nov2010	208.1 (94.6)	37.5 (17.0)	5	73.0 (22.8)	73.2 (22.9)
9Jan2012‡	208.1 (94.6)	37.5 (17.0)	5	74.3 (23.5)	72.2 (22.3)

‡ The ASTM C666 and ASTM C1202 specimens were cast from this batch.

Table 4. G2 batching information.

Placement Date	Grout, lbs (kg)	Water, lbs (kg)	Mix Time, Minutes	Lab Temp., °F (°C)	Grout Temp. After Mixing, °F (°C)
30Nov2010	270.9 (123)	45.3 (20.5)	5, r1, 1	71.4 (21.9)	79.4 (26.3)

Table 5. G3 batching information.

Placement Date	Grout, lbs (kg)	Water, lbs (kg)	Mix Time, Minutes	Lab Temp., °F (°C)	Grout Temp. After Mixing, °F (°C)
7Nov2011	194.4 (88.2)	31.1 (14.1)	5	72.9 (22.7)	73.7 (23.2)

Table 6. T1 batching information.

Placement Date	Grout, lbs (kg)	Water, lbs (kg)	Mix Time, Minutes	Lab Temp., °F (°C)	Grout Temp. After Mixing, °F (°C)
8Dec2010	188 (85.3)	49.7 (22.5)	12.5	70.9 (21.6)	74.4 (23.6)

M1

The batching information for M1 is in Table 7. The guidelines given by the manufacturer were followed as shown in Appendix A.5. This particular mix was designed for ambient temperatures less than 85°F (29.4°C) which was met throughout. The material proportions and mix times were strictly followed due to explicit warnings in the manufacturer's information. Thirteen mixes were used for the initial casting, and three more for the durability test specimen casting. The large number of mixes resulted from the approximately 10 minute working time from the moment mixing began until initial set. The temperature after mixing remained close to the ambient laboratory temperature; however, surface temperatures at final set approximately ten minutes after the completion of mixing and placement were measured to be in excess of 185°F (85°C).

Table 7. M1 batching information.

Placement Date	Grout, lbs (kg)	Water, lbs (kg)	Mix Time, Minutes	Lab Temp., °F (°C)	Grout Temp. After Mixing, °F (°C)
15Dec2010[*]	25 (11.3)	2 (0.91)	1.5	73.6 (23.1)	81.5 (27.5)
9Jan2012[**,‡]	27.8 (12.6)	2.3 (1.04)	2.5	74.1 (23.4)	77.8 (25.4)

* Thirteen mixes were mixed and placed consecutively due to the short work time.
** Three mixes were mixed and placed consecutively due to the short work time.
‡ The ASTM C666 and ASTM C1202 specimens were cast from this batch.

E1

E1 was mixed in the pan mixer but with slightly different procedures as explained in Appendix A.6 and shown in Table 8. A two part premeasured epoxy consisting of a resin and hardener was mixed with a paddle mixer in a plastic bucket for approximately 1 minute. The dry aggregate compound was then placed in the typical pan mixer (Figure 1) and the epoxy compound was added over approximately a 1 minute period following the start of the mixer. The entire batch of E1 (two part epoxy and aggregate) was mixed for an additional 3 minutes at which point it was fluid and ready to be placed.

Table 8. E1 batching information.

Placement Date	Dry Aggregate, lbs (kg)	Resin, lbs (kg)	Hardener, lbs (kg)	Mix Time, Minutes	Lab Temp., °F (°C)	Grout Temp. After Mixing, °F (°C)
22Dec2010	252.4 (114.5)	24.0 (10.9)	4.0 (1.8)	1.0 resin/hardener, 3.0 everything	73.9 (23.3)	74.2 (23.4)
10Jan2012‡	126.2 (57.2)	12.0 (5.44)	2.0 (0.91)	1.0 resin/hardener, 3.0 everything	73.2 (22.9)	76.2 (24.6)

‡ The ASTM C666 and ASTM C1202 specimens were cast from this batch.

U1

The U1 mix design was based on manufacturer recommended proportions. The premix, superplasticizer, and steel fibers were provided by the manufacturer (See Appendix A.7). The mixing procedure included a stepped process of adding premix, mixing in the fluids, reaching a flowable consistency, and then finally adding fibers. The U1 batching information is shown in Table 9.

Table 9. U1 batching information.

Placement Date	Premix, lbs (kg)	Water, lbs (kg)	Super-plasticizer, lbs (kg)	Steel Fibers, lbs (kg)	Mix Time, Minutes	Lab Temp., °F (°C)	Grout Temp. After Mixing, °F (°C)
5Jan2011	137.1 (62.2)	8.1 (3.68)	1.88 (0.85)	19.5 (9.75)	27.5	74.6 (23.7)	78.6 (25.9)
5Jan2011	137.1 (62.2)	8.1 (3.68)	1.88 (0.85)	19.5 (9.75)	25.5	75.2 (24.0)	79.3 (26.3)

Note: Mixer limitations necessitated the sequential mixing of two batches.

U2

As with U1, the U2 mix design was based on manufacturer recommended proportions. The premix, superplasticizers, and steel fibers were provided by the manufacturer (See Appendix A.8). The mixing procedure for U2 was similar to U1 with one key adjustment. Both superplasticizers were stirred into the mixing water before any liquid was added to the dry premix. Aside from this change, mixing included the same stepped process as described for U1 of adding premix, adding the fluids, allowing the resulting material to reach a flowable consistency, and then finally adding fibers. The U2 batching information is shown in Table 10.

Table 10. U2 batching information.

Placement Date	Premix, lbs (kg)	Water, lbs (kg)	Super-plasticizer #1, lbs (kg)	Super-plasticizer #2, lbs (kg)	Steel Fibers, lbs (kg)	Mix Time, Minutes	Lab Temp., °F (°C)	Grout Temp. After Mixing, °F (°C)
7Feb2012	150.7 (68.3)	10.3 (4.67)	1.236 (.561)	0.824 (0.374)	10.7 (4.859)	12.0	74.5 (23.6)	81.2 (27.3)
7Feb2012	150.7 (68.3)	10.3 (4.67)	1.236 (.561)	0.824 (0.374)	10.7 (4.859)	11.5	74.1 (23.4)	81.7 (27.6)
10Jan2012‡	89.1 (40.4)	6.1 (2.76)	0.73 (0.331)	0.49 (0.221)	6.33 (2.871)	11.5	74.3 (23.5)	76.2 (24.6)

Note: Mixer limitations necessitated the sequential mixing of the first two batches listed in the table.
‡ The ASTM C666 and ASTM C1202 specimens were cast from this batch.

C1

C1 (Table 11) was developed based on the published Virginia A4 mix design as shown in Chapter A.9. Multiple trial mixes were produced prior to the final three mixes used in the tests. The mix proportions were consistently close to the mix design, however the average slump was slightly over 5 in. (12.7cm) and the average air content was approximately 1%. A laboratory mixing procedure of 3 minutes mix, 2 minutes rest, and 2 minutes mix was used. The lab temperature stayed at approximately 70°F (21.1°C) throughout. Note that a traditional deck concrete mix design would produce a concrete with significantly higher air content in order to resist freeze-thaw degradation; thus, the deck concrete engaged here did not precisely replicate that which would commonly be deployed in the nation's bridge inventory.

Table 11. C1 batching information.

Placement Date	Cement, lbs (kg)	Water, lbs (kg)	Coarse Aggreg., lbs (kg)	Fine Aggreg., lbs (kg)	Mix Time, Minutes	Lab Temp., °F (°C)	Temp. After Mixing, °F (°C)
1Feb2011*	79.3 (36.0)	35.7 (16.2)	252.0 (114.5)	124.0 (56.4)	3, r2, 2	71.2 (21.7)	71.1 (21.7)

*Three separate batches were made within the testing program.

CONSTRUCTABILITY

Workability

Workability was determined by performing flow measurements on the grouts, obtaining the slump of the concretes, and observing the ease in use of the materials. This included documenting the placement, the cleanup, and the demolding procedures and noting any difficulties.

Immediately after mixing, the flow was measured for every material except C1 (Table 12). A standard slump was taken for C1 according to the procedures described in ASTM C143-10 *Slump of Hydraulic-Cement Concrete*. The flow measurements were based on ASTM C1437-07 *Flow of Hydraulic Cement Mortar*. The spreads of the materials were first computed for the grouts immediately after releasing the grout and prior to dropping the table. Figure 2 shows a typical flow measurement for grout prior to dropping the table. After this measurement, the table was dropped either 25 times (according to ASTM C1437-07) or until the grout flowed off the table indicating a spread greater than 10 in. (25.4 cm).

Figure 2. Photo. Flow measurement of E1.

Table 12. Spread measurements using ASTM C1437-07 methods.

Material	Initial Spread (NO Table Drops), in. (cm)	Final Spread (Number of Table Drops Noted)	
		Table Drops	Spread, in. (cm)
G1	4.8 (12.2)	17	10.0 (25.4)
G2	4.0 (10.2)	9	10.0 (25.4)
G3	4.0 (10.2)	24	10.0 (25.4)
M1	6.6 (16.8)	25	7.6 (19.2)
E1	6.8 (17.3)	25	7.2 (18.2)
T1	10.0 (25.4)	*	*
U1	7.1 (18.0)	25	8.5 (21.7)
U2	10.0 (25.4)	*	*

*T1 and U2 spread 10 in. (25.4 cm) without any table drops.

The three standard grouts, G1, G2, and G3, exhibited full spreads using less than 25 table drops. T1 and U2 flowed off the table without any table drops. These five materials exhibited a fluid consistency and were easy to place. M1, E1, and U1 had spreads between 6.6 in. (16.8 cm) and 7.1 in. (18.0 cm) without dropping the spread table. When the table was dropped 25 times the spread increased by approximately 1.0 inch (2.5 cm) on average for each material. These materials are considered to have exhibited flowable characteristics. All of the grouts were easy to use when pouring them into the specimen molds.

C1 had course aggregate and was not flowable but was very workable. Over the course of using this mix, the average slump was 5.3 in. (13.5 cm). It was easy to place in all specimens aside from the shrinkage rings. These rings were narrow (1.5 in. (3.8 cm) wide) and required extensive rodding to consolidate C1. This exemplified why a fluid grout without course aggregate is desirable in very narrow connections or connections with congested rebar. Table **13** presents notes on the mixing and placing procedures for each material. Aside from M1, all of the materials were workable for at least 30 minutes, the time needed to fabricate all the specimens. M1 was workable for less than 10 minutes on average. A hand held paddle mixer was used to mix M1 inside a five gallon bucket (Figure 3). The number of mixes was substantially larger in order to cast all of the specimens within the shortened work time.

Cleaning tools and mixers was easy with standard grout and concrete mixes but more challenging with other materials. T1 and E1 were very sticky and required abrasion to clean the tools. M1 reached a setting point so quickly that tools had to be cleaned between every mix. U1 and U2 were not hard to clean, but the steel fiber reinforcement contained therein did necessitate modified casting and cleaning procedures.

All the materials were easy to demold except for M1 and E1. Both of these materials bonded very well to the steel forms (Figure 4). Note that these grouts did not bond as well to plastic forms, therefore plastic formwork might be considered when producing material characterization specimens from these grouts.

Table 13. Mixing notes.

Material	Work Time*	Number of Pours†	Cleanup Issues	Demolding Notes
G1	Sufficient	1	Easy	Easy to demold
G2	Sufficient	1	Easy	Easy to demold
G3	Sufficient	1	Easy	Easy to demold
M1	Average of 10 Minutes	13	Hard to clean tools, Clean every pour	Expansive, Bonds well to steel
E1	Sufficient	1	Very sticky, Hard to clean tools	Bonds very well to steel
T1	Sufficient	1	Sticky, Easy cleanup	Easy to demold
U1	Sufficient	2	Bonds to tools, Needle-like fibers	Not set at 24 hours, Easy to demold
U2	Sufficient	2	Bonds to tools, Needle-like fibers	Easy to demold
C1	Sufficient	1	Easy	Easy to demold

* "Sufficient" indicates that there were no problems pouring the specimens (approx. 30 minutes).
† Number of pours used to make all the test specimens.

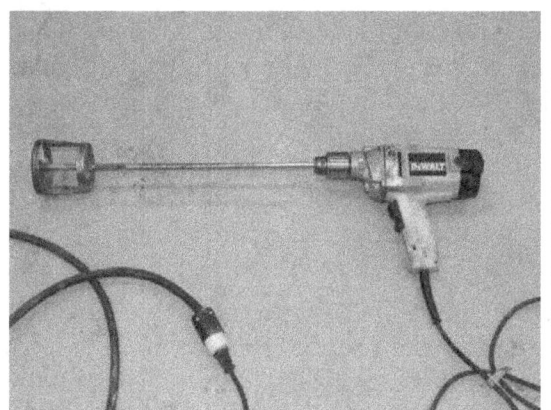 

Figure 3. Photo. The paddle mixer and mixing M1.

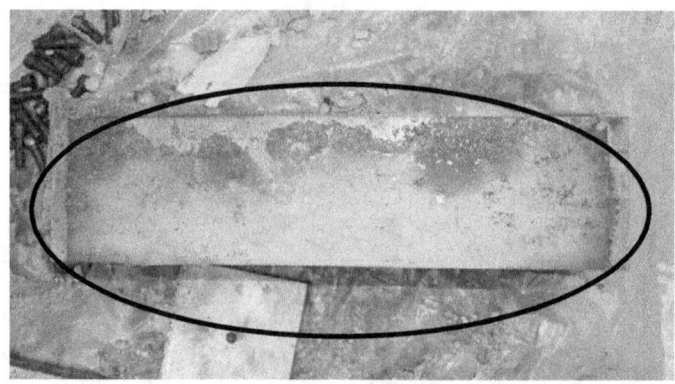

Figure 4. Photo. Cohesion of M1 to steel formwork.

Set Time

The set time was measured based on ASTM C403-08 *Time of Setting of Concrete Mixtures by Penetration Resistance*. The test is based on measuring the pressure needed to force a flat-headed, small-diameter cylinder to penetrate 1 in. (25.4 cm) into the mortar being tested. The mortar is cast into a 6 in. (152 cm) diameter cylinder mold to a depth of 5.5 in. (140 cm). Readings are taken periodically after placing the mortar until a pressure of 4000 psi is surpassed. The data is plotted as pressure versus time after mix initiation. The time to reach 500 psi (3.45 MPa) and 4000 psi (27.6 MPa) of penetration resistance correspond to the initial and final set of the mortar, respectively. Each grout was used as the mortar in this ASTM method to determine how quickly it reached initial and final set as defined by this method.

The ASTM C403-08 tests reported herein were performed on an alternate set of material placements that occurred between April and August of 2011 as part of a separate research project (Swenty, M., and Graybeal, B., 2012). The manufacturers of the materials, mix design, and mix procedures were the same as was described earlier in this report. Curing was performed in the same manner in the concrete materials lab at TFHRC. The samples remained indoors in the lab under burlap and plastic except when penetrometer measurements were captured. The only difference was that these alternate placements were poured approximately 6 months after the first set of placements.

The times for initial and final set were computed using ASTM C403-08 as a guideline. The resistance of the penetrometer and the time after mix initiation were recorded throughout the curing process. A best fit curve was created on the graph of resistance versus time data. The initial and final set points were determined based on the equation of the curve. The initial set was equal to 500 psi (3.4 MPa) resistance and the final set was equal to 4000 psi (27.6 MPa) resistance. A sample graph for G1 is shown in Figure 5. The time to initial and final set is defined as the time from the moment mixing began to the point when the penetrometer read the critical resistance.

Table 14 presents the results for each material. One modification to the ASTM test procedure was that only one sample was taken for each individual material placement due to the mixer volume limitation. Materials with more than one sample correspond to multiple material placements with the same mix design. This number corresponds to the number of tests performed during this alternate research project and has no correlation to the number of samples recommended by ASTM C403-08.

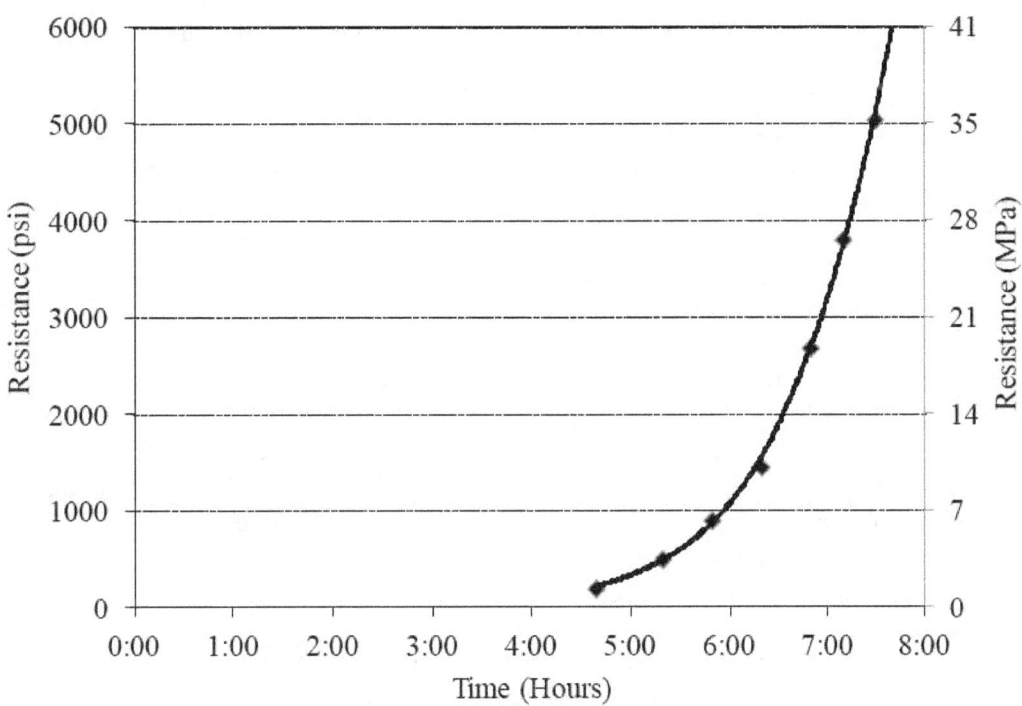

Figure 5. Graph. Typical set time development graph.

Table 14. Penetrometer measurements.

Material	Samples	Initial Set (Hours:Minutes)	Final Set (Hours:Minutes)
G1	9	5:15	6:50
G2	2	8:55	10:25
M1	1	0:07	0:08
E1	1	2:05	2:20
U1	2	8:20	16:55
U2	1	1:10	5:00
C1	2	3:45	6:00

*G3 and T1 were not tested in this phase of the research.

M1 reached initial and final set the quickest. This material set so fast that only two data points were taken at 8 and 10 minutes. The initial set and final set were rounded to the nearest minute at 7 and 8 minutes, respectively. This data indicates M1 sets within minutes and provides little work time between the end of mixing and initial set.

E1 was the second material to reach final set. The time between initial and final set was 15 minutes, indicating a rapid increase in penetration resistance once the curing reaction began.

Grouts G1 and U2 as well as concrete C1 reached final set between 5 and 7 hours after start of mixing. U1 reached final set the slowest in nearly 17 hours. This result confirms that some UHPC formulations exhibit a long dormant period prior to the full initiation of the curing reactions while others set and gain strength more quickly. Note that the U2 grout is specifically designed to set more quickly than U1, with the more rapid strength gain detailed in research published elsewhere (Graybeal, B., and Stone, B., 2012).

Cost

All of the materials used in this study, except for U1 and U2, were obtained from local suppliers in the Washington D.C. metropolitan region. Appropriate quantities of each type of grout were acquired through individual purchases during the timeframes noted in Table 15. The constituent materials for the reinforced concrete were purchased individually. U1 and U2 were purchased directly from the manufacturer. It is recognized that cost differences may occur if larger quantities of materials are purchased or if the materials are purchased from a ready mix supplier. Transportation costs were not included. The costs are shown in Table 15.

The least expensive material was C1. The combined cost of C1 was $178/yd^3 ($233/m^3) for the materials. All of the prebagged grout products were significantly more expensive. The standard grout prices ranged from $845/yd^3 ($1105/m^3) to $1881/yd^3 ($2458/m^3). U1 and M1 were both approximately $2000/yd^3 ($2614/m^3). U2 was slightly more expensive at $2200/yd^3 ($2878/m^3). The most expensive material was E1 with a cost of $4577/yd^3 ($5982/yd^3).

Table 15. Bulk material unit cost.

Material	Unit Cost $/yd^3	($/m^3)	Material Acquisition
G1	1566	(2047)	Fall 2010
G2	1881	(2458)	Fall 2010
G3	845	(1105)	Fall 2011
M1	2077	(2715)	Fall 2010
E1	4577	(5982)	Fall 2010
T1	995	(1300)	Fall 2010
U1	2000	(2614)	Fall 2010
U2	2200	(2878)	Fall 2011
C1	178	(233)	Fall 2010

MATERIAL PROPERTIES

Unit Weight

The unit weight was computed by using ASTM C138-10b *Density (Unit Weight), Yield, and Air Content (Gravimetric) of Concrete* in conjunction with ASTM C231-97 *Air Content of Freshly Mixed Concrete by the Pressure Method*. The unit weight was found by using the "measuring bowl" from the pressure meter. The volume had been previously calibrated for the pressure meter assembly.

The unit weights of the different materials varied from 105.6 lb/ft^3 (1692 kg/m^3) for T1 to 159 lb/ft^3 (2547 kg/m^3) for U1. M1 and E1 were at the midrange at 125.9 lb/ft^3 (2017 kg/m^3) and 133.7 lb/ft^3 (2142 kg/m^3), respectively. C1 expressed a unit weight in the range commonly observed for conventional concrete. Both U1 and U2 were slightly heavier, partially due to the internal steel fiber reinforcement increasing the overall density. The full results are presented in Table 16.

Table 16. Unit weights.

Material	Unit Weight lb/ft^3	(kg/m^3)
G1	119.0	(1906)
G2	143.1	(2292)
G3	111.1	(1780)
M1	125.9	(2017)
E1	133.7	(2142)
T1	105.6	(1692)
U1	159.0	(2547)
U2	154.0	(2511)
C1	150.3	(2408)

Compressive Strength

The compressive strength was measured using ASTM C39-09a *Compressive Strength of Cylindrical Concrete Specimens* for material C1 and ASTM C109-02 *Compressive Strength of Hydraulic Cement Mortars (Using 2-in. or [50-mm] Cube Specimens)* for the other materials. All cylinders were 4 in. (10.2 cm) diameter by 8 in. (20.3 cm) nominal length and all cubes were 2 in. (5.1 cm) on each side. Tests were completed at 7 days, 28 days, and a few other specific timeframes as referenced from the initiation of mixing of each material. The lone exception to this was the 28-day testing of U2 which was completed on 3 in. (7.6 cm) diameter, 6 in. (15.2 cm) long cylinders according to ASTM C39-09a. This cylinder geometry is commonly used when testing UHPC compressive strength. The compressive strength results are listed in Table 17.

The first strength reading was attempted within the first 6 hours of testing if the material could be demolded. As seen in the table, M1 and E1 both had significant strength within 6 hours. M1 exhibited 5.49 ksi (37.9 MPa) of strength at 2 hours. E1 exhibited a strength of 3.28 ksi (22.6 MPa) within 6 hours.

G1, G2, and G3, the three standard grouts, exhibited strengths between 3.45 ksi (23.8 MPa) and 5.07 ksi (35.0 MPa) at 24 hours.

U2, E1, and M1 all exhibited high compressive strengths at 24 hours. U2 and E1 exhibited approximately 10 ksi (69 MPa) of compressive strength at 24 hours. E1 exhibited an 8.4 ksi (58 MPa) compressive strength at 24 hours.

C1, T1, and U1 had very little compressive strength at 24 hours but had achieved 4.04 ksi (27.9 MPa), 5.25 ksi (36.2 MPa), and 15.7 ksi (108 MPa) of strength at 7 days, respectively. The compressive strengths at 28 days ranged from 4.87 ksi (33.6 MPa) for C1 to 21.8 ksi (150 MPa) for U2.

Table 17. Compressive strength results.

Material	Average Compressive Strength, ksi (MPa)				
	2 Hours	6 Hours	24 Hours	7 Days	28 Days
G1	*	*	3.45 (23.8)	6.22 (42.9)	6.70 (46.2)
G2	*	*	5.07 (35.0)	7.90 (54.5)	8.94 (61.6)
G3	*	*	3.91 (27.0)	7.16 (49.4)	7.53 (51.9)
M1	5.49 (37.9)	Not Tested	8.40 (57.9)	8.10 (55.8)	9.91 (68.3)
E1	*	3.28 (22.6)	10.1 (69.6)	14.1 (97.2)	14.4 (99.3)
T1	*	*	*	5.25 (36.2)	8.47 (58.4)
U1	*	*	*	15.7 (108)	18.3 (126)
U2	*	*	$10^{\dagger}$ (68.9)	Not Tested	$21.8^{\ddagger}$ (150)
C1	*	*	1.51 (10.4)	4.04 (27.9)	5.87 (33.6)

*Material had not yet set. †Avg. of (3) 2-in. cubes. ‡Avg. of (6) 3 x 6-in. cylinders.

Split Cylinder Tensile Strength
The splitting tensile strength of the materials was found using ASTM C496-04 *Splitting Tensile Strength of Cylindrical Concrete Specimens*. The cylindrical specimens were all 4 in. (10.2 cm) diameter with a nominal length of 8 in. (20.3 cm). U1 and U2 results are not presented. The splitting tensile strength as reported by the ASTM C496-04 test method is not indicative of the cementitious matrix tensile cracking strength of UHPC due to the presence of a high concentration of fiber reinforcement. Although a modified version of ASTM C496 can be used to capture an indication of the tensile cracking strength of UHPC (Graybeal, 2006), this test was not completed as part of the present study. Table 18 contains the final results for the materials tested.

E1 had the highest splitting tensile strength at 1 and 28 days. The 1 day strength of E1 was over four times stronger than the next highest material, G2. The 28 day strength of E1 was three to four times stronger than the standard grouts, M1, and C1. At 28 days the standard grouts, C1, and M1 ranged from 525 psi (3.62 MPa) to 665 psi (4.59 MPa). T1 had the lowest 28 day strength. All of the grouts had higher 1 day strengths, ranging from 330 psi (2.28 MPa) to 435 psi (3.00 MPa), than the C1 strength of 210 psi (1.45 MPa).

Table 18. Splitting tensile strength results.

Material	Average Splitting Tensile Strength, psi (MPa)	
	1 Day	28 Days
G1	385 (2.65)	525 (3.62)
G2	435 (3.00)	665 (4.59)
M1	330 (2.28)	650 (4.48)
E1	1,940 (13.4)	2,130 (14.7)
T1	350 (2.41)	475 (3.28)
C1	210 (1.45)	570 (3.93)

Modulus of Elasticity

ASTM C469-02 *Static Modulus of Elasticity and Poisson's Ratio of Concrete in Compression* was used as a guide in finding the modulus of elasticity of each material. The longitudinal strain values were obtained using a compressometer with a dial gage. The specimens were each loaded twice and the average strain values were used in the final computations. Standard 4 in. (10.2 cm) diameter by 8 in. (20.3 cm) nominal length specimens were used throughout.

The measured modulus of elasticity values are shown in Table 19. Grouts G1, G2, and E1 were all within the 1,000 - 4,000 ksi (6,900 – 27,600 MPa) range that is typically anticipated for cementitious pastes (Mindess, Young, & Darwin, 2003). T1 was slightly lower than 1,000 ksi (6,900 MPa). This low value was likely caused by extensive surface cracks throughout all of the T1 test cylinders. C1, U1, U2, and M1 expressed results commensurate with the traditional relationship between compressive strength and modulus of elasticity for concrete.

Table 19. Average modulus of elasticity.

Material	Average 28 Day Modulus of Elasticity, ksi (GPa)
G1	2300 (15.9)
G2	3100 (21.4)
M1	4770 (32.9)
E1	3390 (23.3)
T1	730 (5.0)
U1	7550 (52.0)
U2	7370 (50.8)
C1	3940 (27.1)

Restrained Shrinkage

ASTM C1581-09a *Determining Age at Cracking and Induced Tensile Stress Characteristics of Mortar and Concrete under Restrained Shrinkage* was used as a guide to compare the propensity of the materials to crack under restrained shrinkage. Each material was cast inside a controlled environment with a temperature between 75°F $^{+}/-$ 4°F (23.9°C $^{+}/-$ 2.2°C) and 45% $^{+}/-$ 5% humidity. For the first 24 hours the rings were covered with wet burlap and plastic (except for M1). The rings were demolded 24 hours after casting and cracking was monitored visually and with strain gages on the inner ring (Figure 6). Four gages were equally spaced around the inner steel ring; however, some of the gages failed during casting thus leaving only three gages recording valid data on some specimens.

Figure 6. Photo. Typical shrinkage ring.

A typical strain development plot of the inner steel ring is provided in Figure 9. For most materials there is a distinct shrinkage development as demonstrated by a gain in strain in the inner steel ring. As shown occurring in the figure at 2.9 days, a rapid reduction in strain indicates that cracking has occurred. The cracking of the ring can also be confirmed visually on the specimen. Table 20 provides the age at first cracking determined both visually and with the strain rate plot for each material.

ASTM C1581-09a requires the outer PVC formwork to stay in place until 24 hours after casting. At 24 hours the formwork is removed and data collection is officially begun. Because many of the materials in these tests set faster than typical concretes, shrinkage began to occur much earlier than 24 hours in the rings. The data for the first 24 hours is presented to provide an indication of what happens in the rings during its early age. However, it must be realized that the data may have been affected by the outer formwork which was still in place. On many of the plots there is a jump in strain at 24 hours when demolding took place.

The strain rate factor was found by plotting the square root of time versus strain in the inner ring per ASTM C1581-09a. Four measurements were made and then averaged. The slope of this line is shown in the equation in Figure 7.

$$\varepsilon_{net} = \alpha\sqrt{t} + k$$

with:

ε_{net} = Net strain – The difference in strain in the steel rings from demolding through time t.

α = Strain rate factor – Strain rate for each gage on the inner, steel ring (in./in./day$^{1/2}$)

t = Elapsed time starting from demolding the ring through the period of interest (days)

k = Regression constant – Used when fitting a line to the data.

Figure 7. Equation. Net strain in the shrinkage rings.

The stress rate in each test at cracking was measured using the strain rate factor. Table 20 provides the strain rate factor and stress rate computed from the strain data in each ring. Figure 8 contains the equation from ASTM C1581-09a used to find q, the stress rate in each specimen:

$$q = \frac{G\,|\alpha_{avg}|}{2\sqrt{t_r}}$$

with:

q = Stress rate in the ring (psi/day)

G = 10,470,000 psi

α_{avg} = Absolute value of the average strain rate factor (in./in./day$^{1/2}$)

t_r = Elapsed time at cracking or the end of the test, smallest (days)

Figure 8. Equation. Net stress in the shrinkage rings.

The standard grouts, G1, G2 and G3, cracked within 4 days of their cast (Table 20). These materials showed about a 20 microstrain expansion around 12 hours then they began shrinking by 24 hours. Visual cracking and cracking indicated by the strain gages occurred very close together for G1 and G2. The plots display a distinct strain decrease at cracking (Figure 9, Figure 10, and Figure 11).

Table 20. Shrinkage ring results.

Material	Age at First Cracking (Visual), days	Age at First Cracking (Strain Gage), days	Strain Rate Factor, (in./in.)/day$^{1/2}$ {(mm/mm)/day$^{1/2}$}	Stress Rate, psi/day (MPa/day)
G1	2.9	2.9	0.000097	302 (2.1)
G2	2.8	2.5	0.000077	254 (1.8)
G3	7.1	3.6	0.000040	111 (0.77)
M1	Test stopped at 121.5 days		0.000003	1.22 (0.01)
E1	Test stopped at 114.6 days		0.0000001	0.03 (0.00)
T1	0.9	0.9	---	---
U1	71.4	16.4	0.000049	63.4 (0.44)
U2	48	6.3	0.000089	186 (1.28)
C1	23.6	23.1	0.000013	14.6 (0.10)

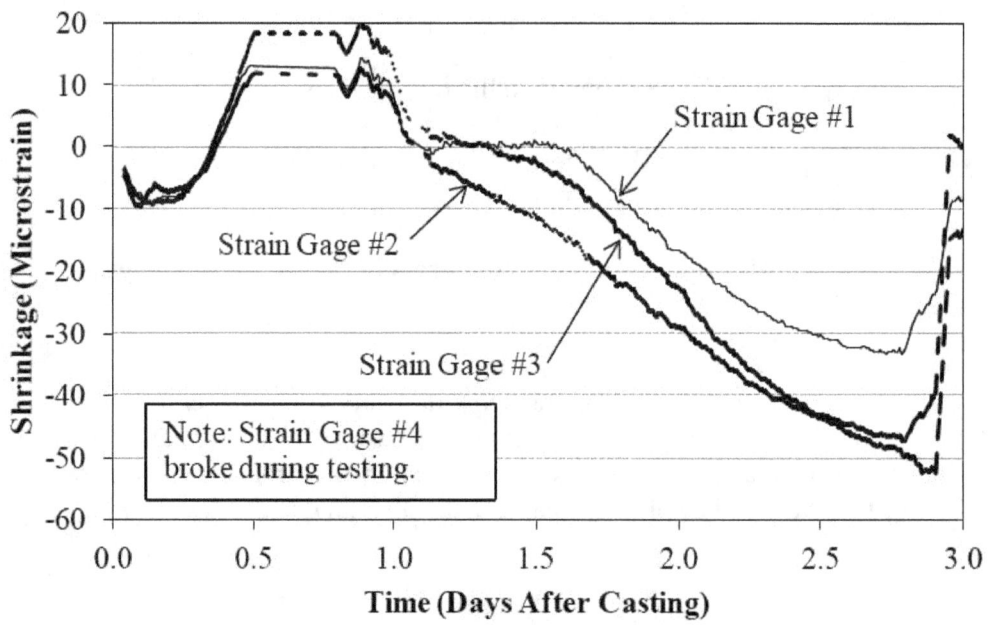

Figure 9. Graph. Strain development in the inner ring over time for G1.

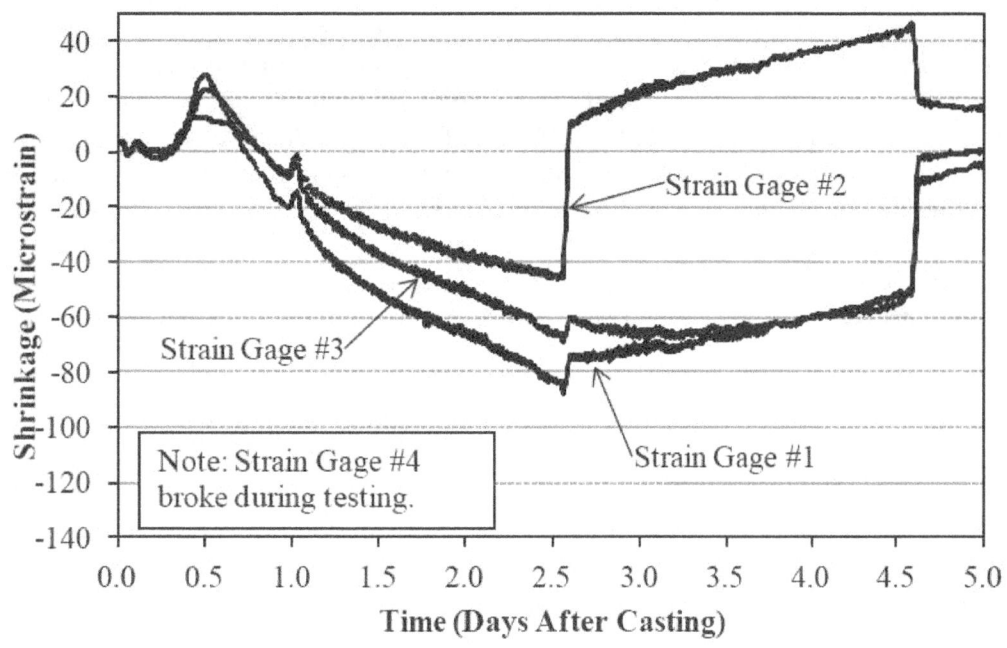

Figure 10. Graph. Strain development in the inner ring over time for G2.

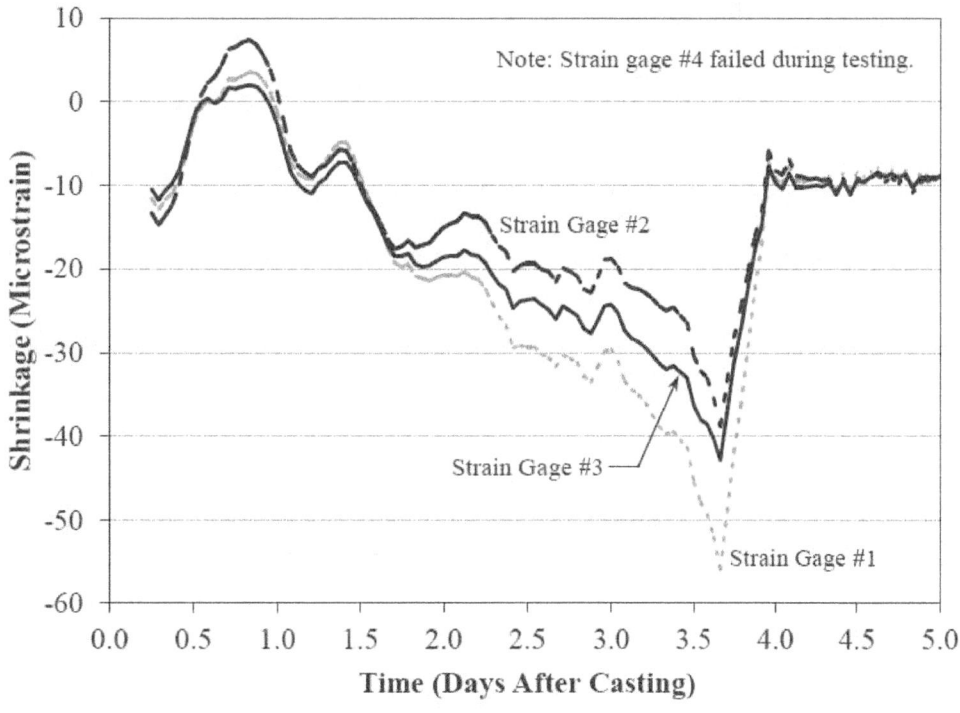

Figure 11. Graph. Strain development in the inner ring over time for G3.

M1 did not crack during testing either visually or as indicated with the strain gages (Figure 12). The ring was monitored for 121 days visually but strain measurements were only taken continuously for the first 30 days. After 30 days strains were measured about every two weeks. During that period a strain rate factor was computed at 0.0000026 (in./in.)/day$^{1/2}$ and a stress rate was computed at 1.22 psi/day (0.01 MPa/day) (Table 20). This rate was computed with the strain readings after demolding (24 hours) through the end of the test. During the first 24 hours the strain gages indicated a large amount of expansion within the grout; however this test method is not meant to measure expansion and may not have provided accurate strain results. M1 does bond well to steel; therefore the expansive readings are likely a close indicator. It must be realized that all the shrinkage in the M1 occurred after initial expansion which likely contributed to the lack of cracking.

E1 demonstrated no cracking either visually or with the strain gage data. An initial shrinkage of approximately 80 microstrain occurred within the first 24 hours (Figure 13). This shrinkage mainly occurred prior to demolding the specimen. After 24 hours there was very little shrinkage in E1. A strain rate factor and stress rate factor of approximately 0.0 (in./in.)/day$^{1/2}$ and 0.0 psi/day (0 MPa/day), respectively, were computed. The initial shrinkage within the first 24 hours was not included in the calculations per the ASTM specification.

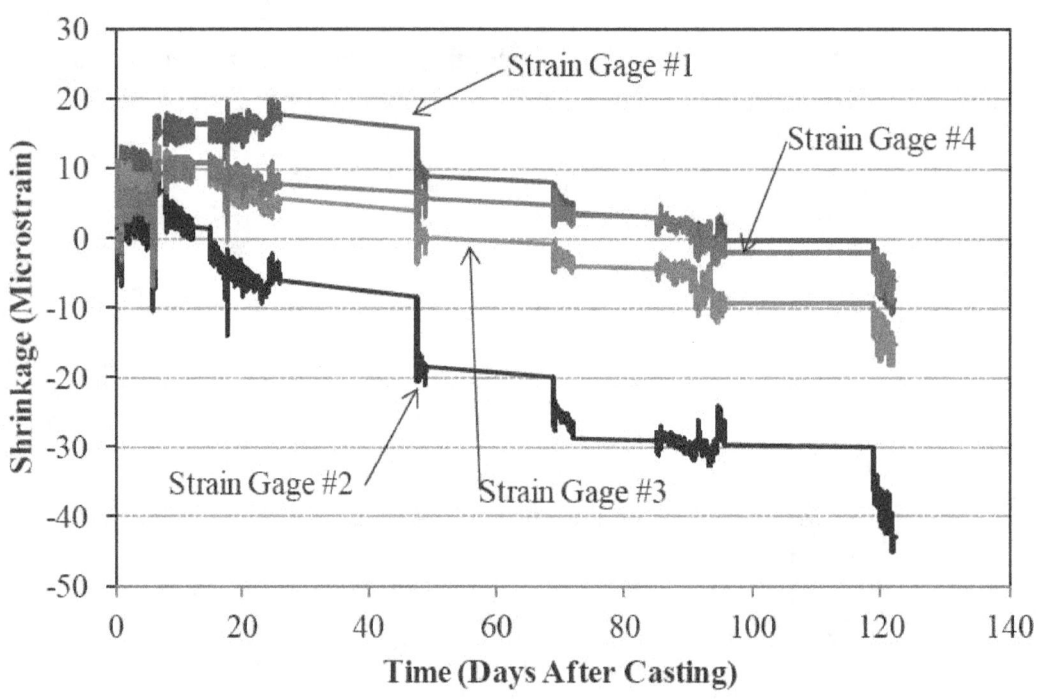

Figure 12. Graph. Strain development in the inner ring over time for M1.

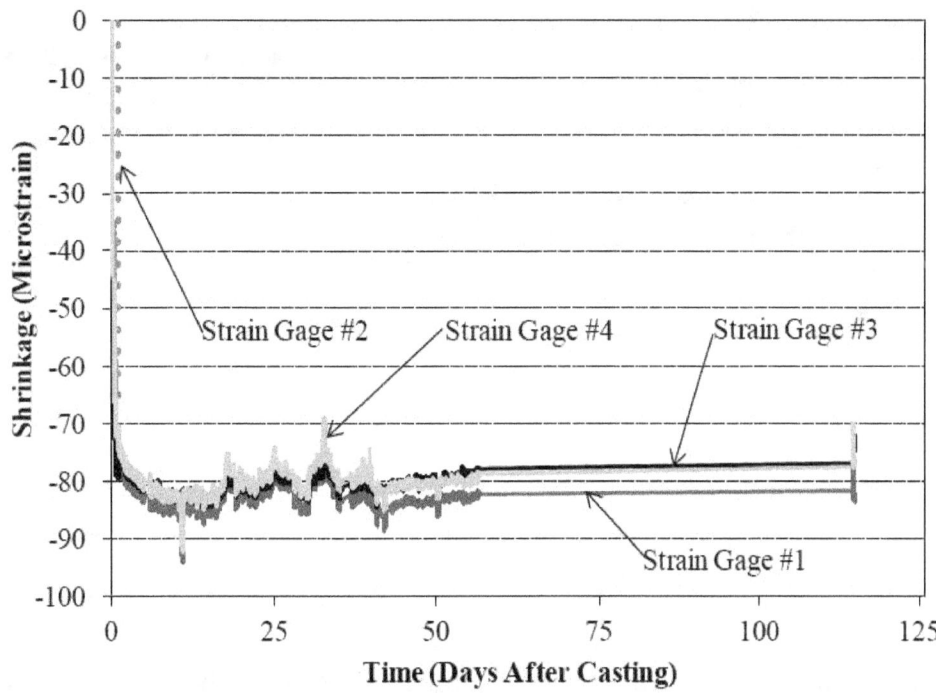

Figure 13. Graph. Strain development in the inner ring over time for E1.

Very little data was collected for T1. The rings were demolded at 24 hours after the pour and cracking was already prevalent through the ring and on its surface. All values for computing the strain and stress rate factors are referenced to the values observed at demolding. Because the rings had already cracked, the test could not be completed according to the ASTM specification. The data that was collected did not indicate any significant strain in the rings prior to demolding.

U1 and U2 behaved differently than typical grouts or deck concretes. Figure 14 and Figure 15 shows the strains in the shrinkage rings versus time. Most notably, these materials never displayed a distinct strain decrease that effectively eliminated the induced strains in the inner steel rings. Instead, they displayed smaller intermittent strain decreases combined with gradual strain increases as the materials continued to hydrate and shrink. This behavior is consistent with steel fiber reinforced cementitious composite materials designed to exhibit post-cracking tensile strength and strain capacity. Additionally, visual identification of cracks in UHPC materials can be difficult, thus making visual identification of first cracking difficult.

C1 cracked later than G1, G2, and G3. Like the grouts, there was an initial expansion of about 10 microstrain during the first 24 hours followed by shrinkage until cracking (Figure 16). The first cracks were detected both visually and electronically on day 23 after casting. The strain rate factor was 0.0000134 (in./in.)/day$^{1/2}$ and the stress rate was 14.6 psi/day (0.10 MPa/day) up until cracking. The loss in strain in the gages was not as distinct as with other materials. There was only an approximately 15 microstrain drop followed by a gradual reduction in strain. Note that an electrical power interruption resulted in intermittent losses of test data between days 18 and 28.

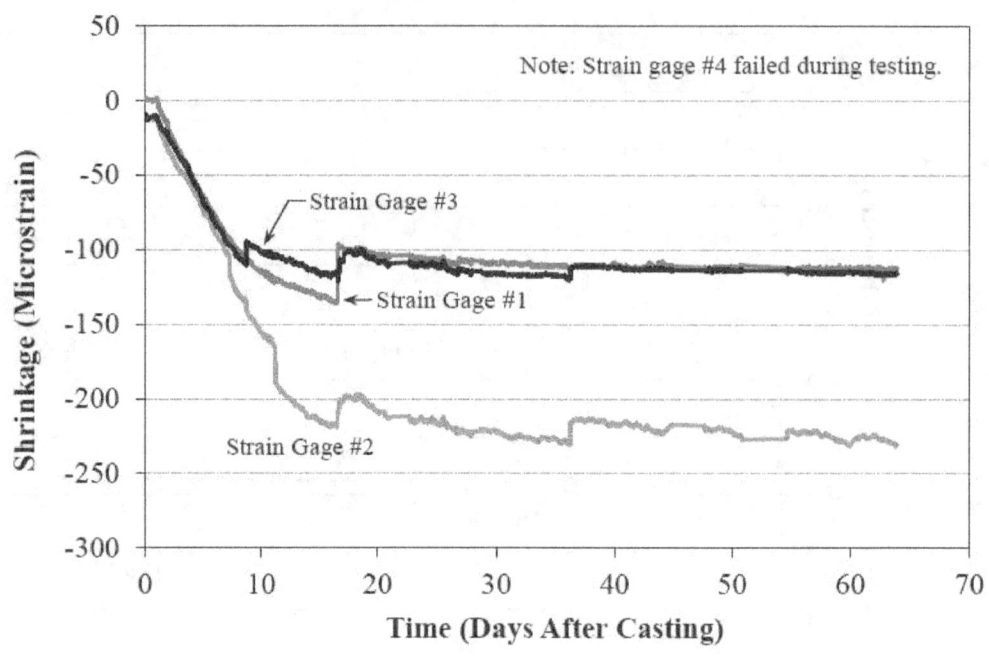

Figure 14. Graph. Strain development in the inner ring over time for the U1.

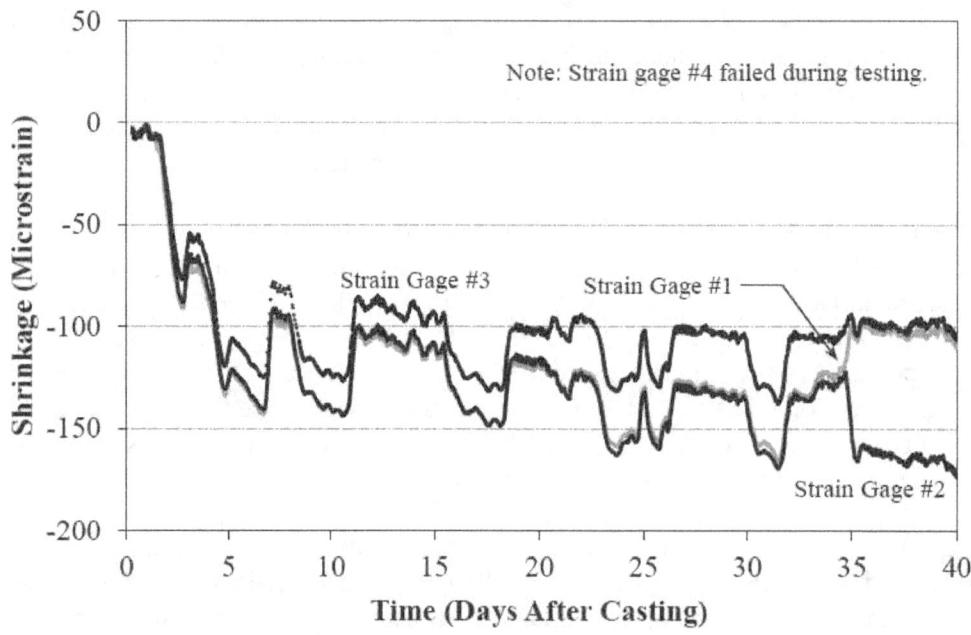

Figure 15. Graph. Strain development in the inner ring over time for the U2.

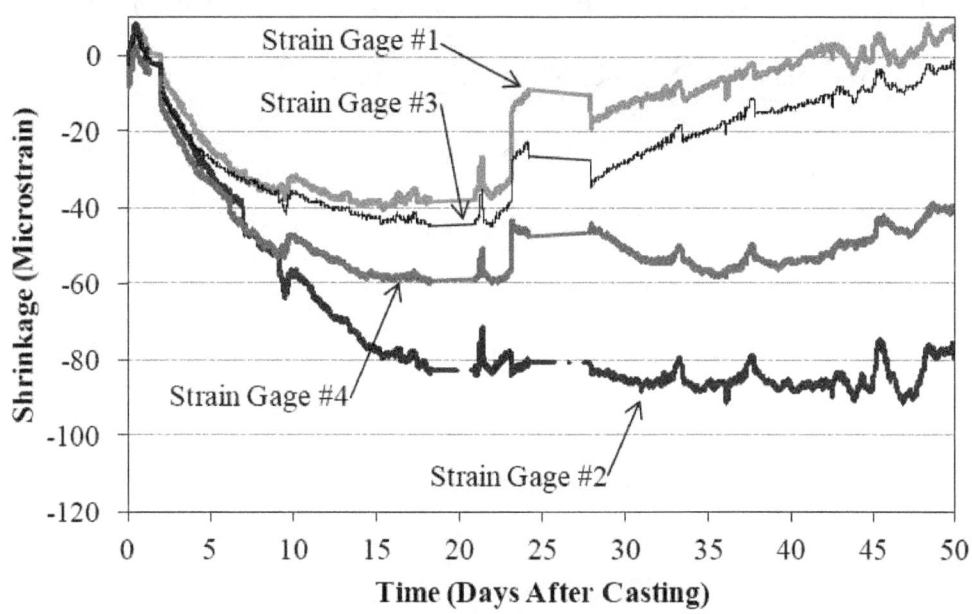

Figure 16. Graph. Strain development in the inner ring over time for C1.

The 28 day crack sizes were recorded for each material (Table 21). M1, E1, U1, and U2 did not have complete vertical cracks visible at 28-days. C1 had two cracks on opposite sides of the ring but the size of each crack was too small to measure with a concrete crack card. The two standard grouts had cracks that were approximately 0.1 in (3 mm) in width at 28 days. T1 was cracked upon demolding and the crack width continued to grow to a width of 0.89 in. (23 mm) at 28 days.

These sizes indicated that the materials with the higher shrinkage values from the ASTM C157-08 tests (discussed in the next section) had larger comparative crack sizes in the shrinkage rings at 28 days. T1 had the highest shrinkage value followed by the standard grouts. These three materials had the largest crack sizes in the shrinkage rings. C1 had similar shrinkage values to E1; however C1 did have two very small cracks. This may have been due to the difference in tensile strengths of the two materials. U1 and U2 clearly displayed shrinkage and behaviors indicative of shrinkage cracking however cracking was not visibly detected at this age. This is likely due to the fiber reinforcement in these materials having arrested small shrinkage cracks, thus preventing these cracks from growing larger and becoming visible to the naked eye.

Table 21. Crack sizes visually observed on shrinkage ring specimens at 28-days.

Material	Number of complete cracks	Size, in. (mm)
G1	1	0.2 (5)
G2	2	0.12 (3)
G3	1	0.12 (3)
M1	None	---
E1	None	---
T1	1	0.89 (23)
U1	None	---
U2	None	---
C1	2	*

*Cracks were smaller than 0.005 in. (0.127 mm).

Unrestrained Shrinkage

Shrinkage was measured with two different methods for every material. The first method was the ASTM C157-08 unrestrained 3 in. by 3 in. by 11 in. (76.2 mm by 76.2 mm by 279.4 mm) shrinkage bars starting 24 hours after the pour. The second method used the same ASTM unrestrained shrinkage bar with an embedded vibrating wire gage (VWG). The VWG method captured unrestrained length change beginning immediately after casting, thus capturing behaviors during the first 24 hours which are not captured in the standard ASTM C157-08 test method.

Early Age Unrestrained Shrinkage

The total shrinkage of a field-cast grout from mixing through the acquisition of full mechanical and durability properties provides an indication of the likelihood that the grout will exhibit shrinkage cracks. Given that the ASTM C157-08 unrestrained shrinkage test only captures shrinkage beginning 24 hours after casting, a non-standardized method was employed to measure early age unrestrained shrinkage including during and after the first 24 hours. Unrestrained shrinkage specimens were made with the same procedures as ASTM C157-08 shrinkage bars with a few exceptions. First, the shrinkage bars were not exposed to a lime bath but were rather left in the environmental chamber for their entire life. This testing program focused on immediate volume change that may occur in a field application with only minimal curing for the first 24 hours. Second, a 6 in. (15.1 cm) long vibrating wire gauge (VWG) was placed directly in the middle of the standard ASTM C157-08 mold (Figure 17). VWGs provide strain measurements along their length in the material they are cast into. Third, the forms were heavily oiled with a form release agent immediately prior to casting the specimens to ensure very little friction developed between the material and form. As such, the specimens were considered to have been unrestrained from casting, through demolding at 24 hours, and to the cessation of data collection. The gauges provided shrinkage measurements to the nearest microstrain and the results were corrected to account for temperature induced dimensional changes.

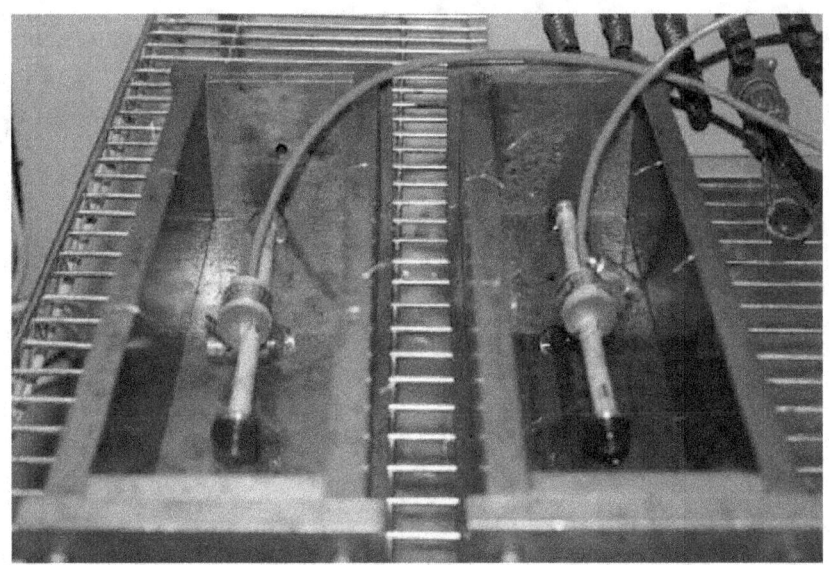

Figure 17. Photo. ASTM C157-08 molds embedded with vibrating wire gages to measure shrinkage.

The results from the shrinkage bars with the VWGs are shown in Figure 18. The shrinkage values are plotted and reported as positive, while the expansion values are shown as negative. Note that shrinkage is plotted with time zero coinciding with the initiation of mixing of the grout material. Shrinkage results are plotted for the eight different materials from the time of pouring through 28 days. The test results for U2 were not captured correctly due to a data acquisition failure, and thus are not presented.

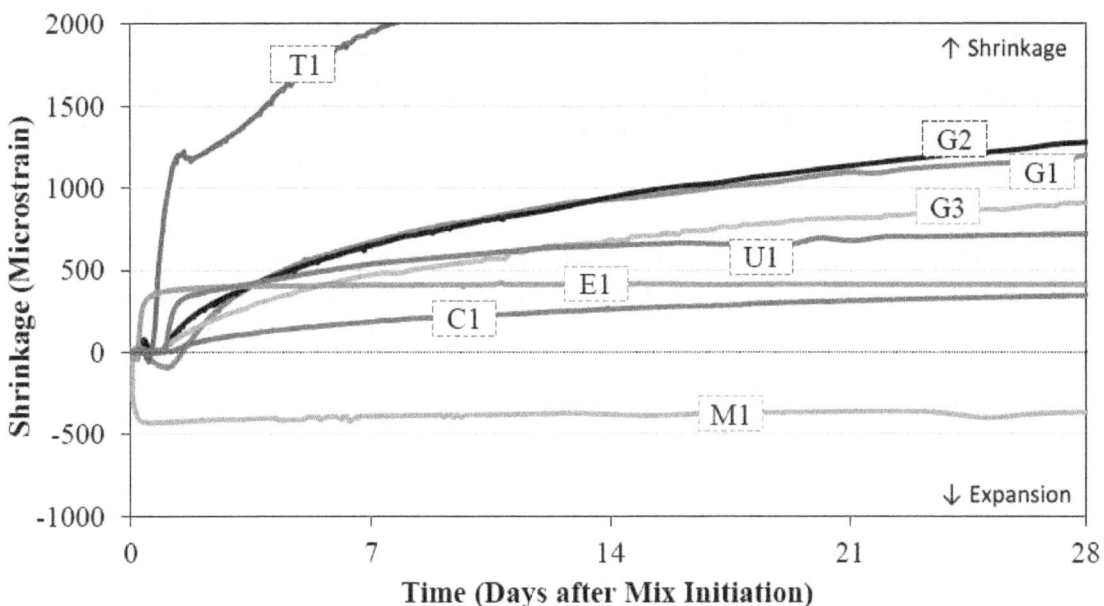

Figure 18. Graph. Unrestrained length change measured via vibrating wire gage.

After one day, material M1 had expanded approximately 400 microstrain and material E1 had shrank approximately 400 microstrain. These values did not change significantly after the initial 24 hours. Note that the test was not designed to allow for unrestrained expansion of a test specimen, thus the expansion of M1 may have been inhibited by the formwork. As such, the M1 may exhibit a net expansion larger than what was observed in this study.

T1 shrank significantly more than the other materials. T1 shrank over 800 microstrain at 24 hours and over 4000 microstrain by 28 days. Grouts G1 and G2 exhibited approximately 1200 microstrain of shrinkage at 28 days while U1 exhibited approximately 700 microstrain at 28 days. The conventional concrete, C1, exhibited less than 400 microstrain of shrinkage at 28 days.

Long Term Unrestrained Shrinkage

Unrestrained shrinkage tests were also completed according to ASTM C157-08 on 3 in. by 3 in. by 11 in. (76.2 mm by 76.2 mm by 279.4 mm) prisms. In order to more closely simulate the field conditions to which these grouts are subjected and to allow for closer comparison with the early age unrestrained shrinkage results discussed previously, the test procedure was slightly modified. Specifically, the ASTM test states that the specimens should be cured in a lime bath during the first 28 days of testing. This testing program focused on immediate volume change that may occur in a field application with only minimal curing for the first 24 hours. With this in mind, the shrinkage bars were not exposed to the lime bath but were rather demolded at 24 hours, measured, and held in a 75°F $^+/-$ 4°F (23.9°C $^+/-$ 2.2°C) temperature and 45% $^+/-$ 5% humidity laboratory environment for the duration of testing.

The results are shown in Figure 19. The shrinkage values are plotted and reported as positive, while the expansion values are shown as negative. Note that, according to the test procedure, the initial reading is captured at 24 hours after initiation of mix initiation. As such, dimensional changes occurring during the first 24 hours are not captured by the test method. Due to the delayed setting of grout U1, the initial reading for this grout was not captured until 48 hours after mix initiation.

According to this test method, grouts E1, M1, and U1 exhibited less than 400 microstrain of shrinkage at 90 days, and concrete C1 exhibited approximately 700 microstrain at 90 days. Material G2 exhibited approximately 1700 microstrain at 90 days. T1 exhibited very high shrinkage values. Note that ASTM C157 shrinkage results were not captured for material G1 using 3 in. prismatic bars; however, ASTM C157 shrinkage testing using 1 in. by 1 in. by 11 in. (25.4 mm by 25.4 mm by 279.4 mm) prisms resulted in an observed shrinkage of approximately 2000 microstrain at 90 days.

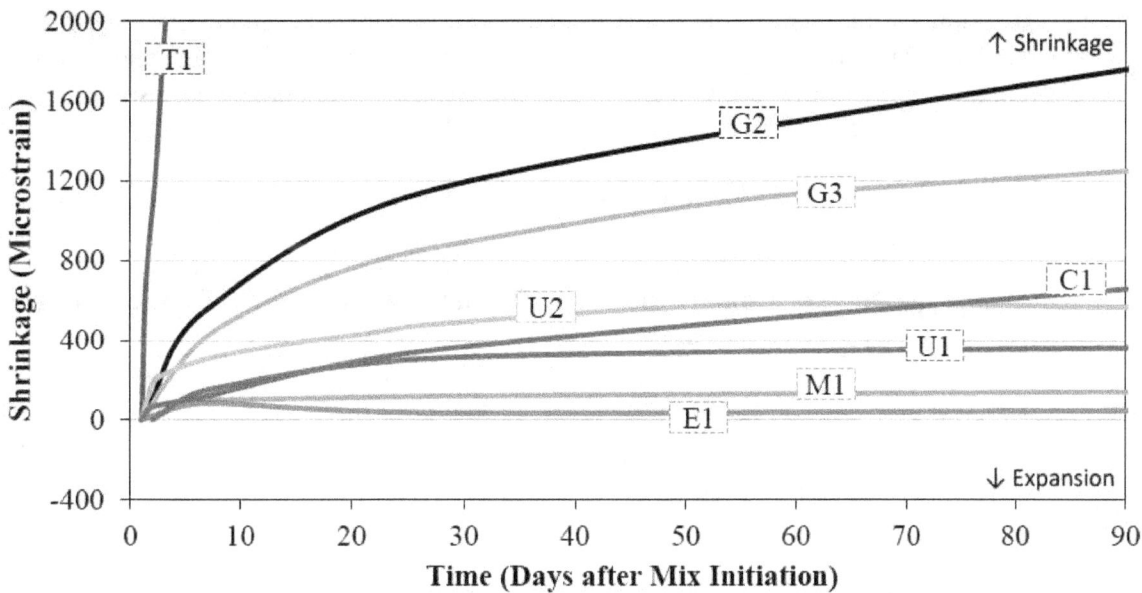

Figure 19. Graph. Unrestrained length change measured via the ASTM C157-08 test method.

BOND TESTING

Three tests were used as an indication of the bond strength between each field-cast grout material and a previously cast and cured deck concrete. The same Virginia A4 bridge deck concrete mix design discussed elsewhere in this study was used here for the precast concrete portion of each test specimen. The concrete for the precast portion of each test specimen was cast approximately two months prior to the placement of the field-cast grouts into the specimen molds. This timing scheme allowed the precast concrete to be nearly dimensionally stable and of appropriate strength prior to the placement of the grout. Two separate batches of the deck mix were cast for the three different bond tests.

The three bond tests engaged within this study included the slant cylinder compression test, the splitting tensile bond test, and restrained shrinkage bond test. The slant cylinder halves and restrained shrinkage bond half rings were made in the first placement of the deck mix, and the splitting tensile bond cylinder halves were made in the second placement. In addition, three 3 in. (7.6 cm) by 3 in. (7.6 cm) by 11 in. (27.9 cm) shrinkage bars and nine 4 in. (10.2 cm) by 8 in. (20.3 cm) cylinders were made with each placement. The ASTM C39-09a compressive strength and ASTM C496-04 splitting tensile strength of the concrete was measured at 7 and 28 days after casting. ASTM C157-08 unrestrained shrinkage behavior was also measured approximately every three days for the first two months then weekly for the third month beginning 24 hours after casting.

The precast concrete properties are shown in Table 22. Mixes from the first two placements had approximately 4,500 psi (31 MPa) compressive strength at 7 days and 5,500 psi (38 MPa) 28-day strength. The objective of producing a generic concrete that might reasonably be used in bridge deck applications with at least a minimum strength of 4,000 psi (27.6 MPa) was met. The corresponding splitting tensile strengths at 28 days for the first and second placements were 600 psi (4.1 MPa) and 680 psi (4.7 MPa), respectively. The same mix design and mixing procedures were used for placement #3, so it is not clear why the compressive strength was in excess of 8000 psi (55 MPa).

Table 22. Precast concrete properties.

Placement	7-Day Compressive Strength,	28-Day Compressive Strength,	28-Day Splitting Tensile Strength, psi (MPa)
#1[†]	4580 (31.6)	5400 (37.2)	600 (4.1)
#2[†]	4460 (30.8)	5590 (38.5)	680 (4.7)
#3[*]	-	8610 (59.3)	-

[†]Placement used with materials G1, G2, M1, E1, T1, U1, and C1
[*]Placement used with materials G3 and U2

The precast concrete was allowed to cure for at least 56 days to allow for significant strength gain and shrinkage to have occurred prior to the casting of the secondary grout material. The shrinkage of the precast concrete was monitored via ASTM C157-08 until 90 days after casting to ensure that the precast concrete was nearly dimensionally stable by the time the field-cast grouts were cast.

Slant Cylinder Compression Test
The first bond test was based on the standard ASTM C882-05 *Bond Strength of Epoxy-Resin Systems Used with Concrete by Slant Shear*. In this standard method, an epoxy-resin-base is bonded between a hardened portland-cement based concrete with a hardened or fresh portland-cement concrete. The intent of the test is to assess the bond performance of the epoxy-resin-base to the hardened or fresh-cast concrete. For the purposes of this testing program, the epoxy resin-base was excluded. The first layer of concrete was the precast deck concrete mix design and the second layer was one of the field-cast grout materials from the testing matrix. The ASTM test specification recommends using 3 in. (7.6 cm) by 6 in. (15.3 cm) cylinder molds, however 4 in. (10.2 cm) by 8 in. (20.3 cm) cylinder molds were chosen for these tests. These larger molds provided a larger surface area for bonding. The slant shear surface was designed based on a 30 degree slant shear plane as measured from the long axis of the cylinder. The precast concrete side was poured using modified plastic cylinder molds placed in special wooden formwork designed to hold the slanted specimens (Figure 20). The concrete was placed in the cylinders in one layer, rodded twenty-five times, and cured with burlap and plastic for 24 hours. The specimens were then placed in the temperature and humidity controlled room until the secondary casting procedures for the field-cast grouts were initiated.

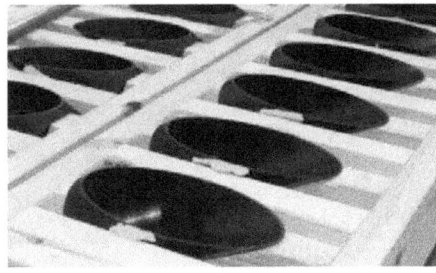 

Figure 20. Photo. Slant cylinder molds and deck concrete halves being sandblasted.

Three slant cylinder samples were made of each material during the testing program. The deck concrete halves were sandblasted approximately 24 hours before casting the second half (Figure 20). Sandblasting was performed with a medium grit (20-40 mesh size) sandblasting media. After sandblasting the bonding surfaces the specimens were covered in saturated burlap and plastic and returned to the environmental controlled room. Approximately an hour before casting, the samples were placed in 4 in. (10.2 cm) by 8 in. (20.3 cm) plastic cylinder molds and left under burlap in the casting room (Figure 21). Once mixing began, the burlap was removed and the field-cast grout material was placed in the mold in 3 equal layers by height. The layers were rodded 25 times with a 3/8 in. (9 mm) diameter rod. The samples were then cured with the same method as all the other samples: 24 hours under burlap and plastic and then in the environmental room until testing. Immediately prior to testing, the ends of the specimens were sulfur capped. The testing procedure followed the steps in ASTM C882-05 which includes an identical loading procedure as in ASTM C39-09a. This was repeated for each type of material.

All tests were conducted 28 days after the placement of the field-cast portion of each test specimen. The bonded surface area was found by measuring the two diameters of the ellipse and computing the area. The maximum load applied was divided by the computed surface area and recorded as bond strength.

Prior to testing G1, G2, T1, and C1, visible cracking appeared at the bonded surface between the two materials. These four materials all had average slant cylinder bond strengths between 200 psi (1.4 MPa) and 920 psi (6.3 MPa) (Table 23).

Difficulty with the casting of the M1 test specimens likely led to the observation of low slant cylinder bond strengths. One of the specimens failed at the interface during demolding, while the other two failed at slant cylinder bond strengths of less than 100 psi (0.6 MPa). During casting, these specimens were not cast quickly enough, likely leading to the initiation of set of the grout prior to stable contact with the precast concrete surface. Given the poor quality of the results for M1 in this test, the results are not provided in Table 23.

The highest bond strengths were observed from E1, U1, and U2. These materials exhibited average slant cylinder bond strengths in excess of 2000 psi (14 MPa). In these samples, the precast concrete substrates broke before or at the same time as the interface bond. Many of the samples broke at loads near the ultimate compressive strength of the precast concrete, indicating that the limiting strength was not the interface bond strength. For these materials, higher bond strengths might be achieved if a higher strength substrate concrete is used.

Table 23. Slant cylinder bond strength.

Material	28-Day Measurements		
	Average Interface Bond Stress at Failure, psi (MPa)	Axial Compressive Stress at Failure, psi (MPa)	Failure Surface
G1	200 (1.4)	390 (2.7)	Along Interface (Precracked)
G2	520 (3.6)	1030 (7.2)	Along Interface (Precracked)
G3*	240 (1.7)	470 (3.2)	†
E1	3530 (24.3)	6830 (47.0)	Through Concrete/Interface
T1	920 (6.3)	1830 (12.6)	Along Interface (Precracked)
U1	2700 (18.6)	5320 (36.7)	Through Concrete/Interface
U2	2200 (15.2)	4390 (30.3)	†
C1	680 (4.7)	1330 (9.2)	Along Interface (Precracked)

*Average calculated from two specimens; 3rd broke along the bond prior to test
†Result not appropriately documented

Splitting Tensile Bond Test

The second bond test was based on ASTM C496-04 *Splitting Tensile Strength of Cylindrical Concrete Specimens*. The standard test method was modified to test the bond strength between two materials rather than the tensile strength of only one material. The specimens consisted of two equal sized halves bonded together lengthwise. A 6 in. (15.2 cm) diameter by 12 in. (30.5 cm) long cylinder was chosen because it provided a large bonded surface between the materials. The first half of the specimens were cast with the deck concrete mix design. Plastic 6 in. (15.2 cm) by 12 in. (30.5 cm) molds were cut in half along the 12 in. (30.5 cm) length and placed in wooden formwork for support (Figure 21). The concrete was poured in one layer and was rodded 36 times with a 5/8 in. (16 mm) diameter rod. The specimens were covered with burlap and plastic for the first 24 hours then demolded and placed in a temperature and humidity controlled room until bonded.

Three split cylinder bond samples were made of each material during the testing program. The deck concrete halves were sandblasted approximately 24 hours before casting the second half (similar to the slant cylinders in Figure 20). Sandblasting was performed with a medium grit (20-40 mesh size) sandblast media. After sandblasting the bonding surfaces, the samples were covered in saturated burlap and plastic then returned to the environmental control room. Approximately an hour prior to casting the second half of the samples, the first halves were placed in 6 in. (15.2 cm) by 12 in. (30.5 cm) molds.

Figure 21 shows the precast pieces in the cylinders just prior to the casting of the field-cast materials. The field-cast materials were placed in the mold in 3 equal depth layers. Each layer was rodded 25 times with a 5/8 in. (16 mm) diameter rod. The samples were then cured with the same method as the other samples: 24 hours under burlap and plastic and then in the environmental room until testing.

The testing procedures mimicked the procedure provided in ASTM C496-04. The bonded plane was aligned perpendicular to the loading surfaces. An alignment device was used to place the wooden 1/8 in. (3 mm) by 1 in. (25 mm) plywood strips directly above the bonded plane and beneath the center of thrust (Figure 22).

Figure 21. Photo. Splitting cylinder bond forms for the first halves; splitting cylinder and slant cylinder forms for the second halves.

Figure 22. Photo. Typical 6 in. by 12 in. split cylinder bond specimen and testing setup.

All tests were conducted 28 days after the placement of the field-cast grout material. The bonded surface area was rectangular in shape based on the diameter and length of the bonded cylinders. The diameter was measured in three locations and the length was measured in two locations for each specimen. The splitting tensile strength was based upon the equation from ASTM C496-04:

$$T = \frac{2P}{\pi LD}$$

with:
- P = Maximum applied load
- L = Average length along the length of the bonded surface
- D = Average diameter along the length of the bonded surface

Figure 23. Equation. Splitting tensile strength from ASTM C496-04.

Prior to testing G1, G2, G3, T1, and M1, small visible cracks appeared at the bonded surface between the two materials. These four materials all had average split cylinder bond strengths between 260 psi (1.79 MPa) and 368 psi (2.54 MPa). As shown in Table 24, these specimens all broke at the bonded interface between the two materials.

C1 did not exhibit visible interface cracks at the bonded surface prior to testing; however, the split cylinder bond strength was 248 psi (1.71 MPa). This was the lowest value of any material bonded to the precast deck concrete.

Table 24. Splitting tensile bond strengths.

Material	Average Splitting Tensile Bond Strength, psi (MPa)	Failure Surface
G1	260 (1.79)	Along Interface (Precracked)
G2	368 (2.54)	Along Interface (Precracked)
G3	300 (2.07)	Along Interface (Precracked)
M1	362 (2.50)	Along Interface (Precracked)
E1	483 (3.33)	Precast Concrete Paste
T1	277 (1.91)	Along Interface (Precracked)
U1	664 (4.58)	Precast Concrete Paste
U2	619 (4.27)	Precast Concrete Paste
C1	248 (1.71)	Along Interface

The highest bonded values came from U1, U2, and E1. U1 and U2 exhibited average splitting tensile bond strengths greater than 600 psi (4.1 MPa). The failure surface in the U1, U2, and E1 materials was located in the precast concrete paste next to the bonded surface. These field-cast grout materials did not fail; rather, the inherent tensile strength of the precast concrete substrate proved to be the limiting factor.

The splitting tensile bond strengths of the materials with deck concrete were compared to the standard ASTM C496-04 splitting tensile strength of the bonded materials. All of the materials' splitting tensile strengths were measured on the same day as the bond tests. As shown in Table 25, most tested materials exhibited splitting tensile bond values less than 60% of their respective ASTM C496-04 splitting tensile strengths. Thus, the bond strength of these field-cast materials to the precast concrete with the tested interface surface preparation was approximately half the strength of a monolithic grout sample without a bonded interface.

A similar comparison was made between the materials' split cylinder bond values (Table 24) and the ASTM C496-04 splitting tensile strength of the precast deck concrete (Table 22). The results are also provided in Table 25. All tested materials except U1 and E1 had bond split cylinder values between 38% and 58% of the substrate concrete ASTM C496-04 strength. The results indicate the bond strength of many grout materials with the precast concrete and the tested interface surface preparation was no more than about half the splitting tensile strength of the precast concrete.

Table 25. Comparison of splitting cylinder and splitting cylinder bond strengths.

Material	Average ASTM C496 Splitting Tensile Strength, psi (MPa)	Bond Strength as a Percent of Field-Cast Material ASTM C496 Splitting Tensile Strength	Bond Strength as a Percent of Precast Concrete ASTM C496 Splitting Tensile Strength
G1	525 (3.62)	49.5	40.7
G2	665 (4.59)	55.3	57.6
G3	730 (5.03)	41.1	†
M1	650 (4.48)	55.7	56.7
E1	2125 (14.7)	22.7	75.6
T1	475 (3.28)	58.3	43.4
U1	*	*	103.9
U2	*	*	†
C1	570 (3.93)	43.5	38.8

† Splitting tensile test on associated precast concrete was not completed.
* The ASTM C496 test method does not report appropriate tensile strength values for UHPC-class materials.

Restrained Shrinkage Bond Test
The third bond test was based on the restrained shrinkage ring test (ASTM C1581-09a). The standard test method was modified to allow the bond between a precast concrete and a field-cast grout to be assessed. Specifically, the ring was created out of two different materials joined at two vertical surfaces. The precast and field-cast portions of the test specimen were symmetric. The first half ring was cast using the concrete bridge deck mix design. The same setup as ASTM C1581-09a was used: a concentric ½ in. (1.3 cm) thickness steel ring and a 1 in. (2.5 cm) thickness PVC ring with a 1 ½ in. (3.8 cm) gap between them. The outside diameter of the steel ring was 13.0 in. (33.0 cm). A wooden blockout was placed in the middle of the forms during the placement of the precast concrete half to create the bonding interface (Figure 24). The concrete was mixed and placed in the concrete materials lab then covered with burlap and plastic

for the first 24 hours. Next, each half specimen was demolded then left uncovered to cure until the casting of the second half of each specimen. The first material, G1, was cast approximately two months after the initial concrete was cast. At this point the unrestrained shrinkage had slowed and the rings were ready to be bonded to another material.

Figure 24. Photo. Ring setup for the placement of the deck concrete half.

The second half of the rings were cast alongside the other test specimens which were cast for each studied material. The bonding interfaces of the precast deck concrete half-rings were sandblasted 24 hours prior to the material pour. Sandblasting was performed with a medium grit (20-40 mesh size) sandblast media. After sandblasting the bonding surfaces the specimens were returned to their original formwork in the environmental room (Figure 25). The rings were aligned to ensure the proper width according to the ASTM test method and then they were affixed to the base plate to ensure that the rings did not move during casting. A saturated piece of burlap was placed over the bonded surface and then covered in plastic. Both were removed right before the placing of the field-cast material. The geometry of the bonded surface (approximately 6 in. (15 cm) tall by 1 ½ in. (3.8 cm) wide) was recorded for each test specimen.

Each material was cast inside a controlled environment with a temperature between 75 $^{+}/-$ 4 degrees Fahrenheit and 45% $^{+}/-$ 5% humidity and left there for the duration of the test. After the field-cast materials were placed, the second half of the rings was cured with wet burlap and plastic for 24 hours. The rings were demolded 24 hours after this casting, and cracking was monitored visually and with strain gages on the inner ring (Figure 26). Monitoring with strain gages began when placing the field-cast material and continued until initial cracking had occurred or at least 100 days had elapsed. Four gages were equally spaced around the inner steel ring; however some of the gages broke during casting leaving only three gages to capture the response of some test specimens. Visual monitoring was also performed every day for the first week and then approximately twice a week for the remainder of each test. Special attention was placed on the interface between the materials to determine if the first cracks occurred at an interface or within one of the half rings.

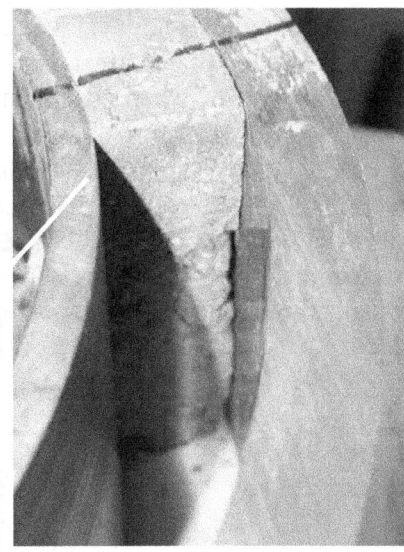

Figure 25. Photo. Ring setup for the placement of the second half.

The results from the strain gage measurements required careful interpretation. Many of the materials did not produce the consistent strain patterns normally observed in ASTM C1581-09a test results. The bonded rings tended to crack quicker than would be expected from a monolithic cast of either material. Smaller strain readings were observed in the steel rings prior to cracking, thus the plots of strain versus time provide less clarity of behavior. This difference of behavior was likely due to the modification of the test method. The ASTM C1581-09a test method assumes that the cast material is shrinking around the circumference of the ring. In general, the bonded rings had one or two strain gages that reported lesser strain values. This may have been due to an uneven strain development around the circumference of the steel ring. Also, the portion of each ring adjacent to the precast concrete did not have as tight a bond to the steel ring due to this half-ring being removed for sandblasting then re-inserted in the ring setup and thus may have impacted the observed results.

Figure 26. Photo. Restrained shrinkage bond test setup.

To overcome these limitations, the date of visual cracking was used as a guideline when interpreting the cracking strain data. The results are presented in Table 26. The graphs were scrutinized at the time when visual cracking was documented in at least one location over the entire height (6 in. (15.2 cm)) of the test specimen. It is assumed that the strain gages would indicate cracking at a time similar to or before the first visual crack appeared. Shrinkage data was recorded from the time of placing the second material until cracking. The data is plotted for the entire period, however it should be noted that during the first 24 hours the rings were enclosed in the formwork. The behavior prior to demolding was influenced by the outer ring, while the demolding at 24 hours may have also caused additional strain in the plots.

Table 26. Restrained shrinkage bond test results.

Material	Age at First Cracking – Visual, Days	Age at First Cracking – Strain Gages, Days	Crack Locations*	28 Day Crack Sizes, in. (cm)	
				Location 1	Location 2
G1	2.9	1.6	2 Interfaces	0.017 (0.043)	0.009 (0.022)
G2	2.0	1.5	2 Interfaces	0.009 (0.022)	0.009 (0.022)
G3	7.1	1.5	1 Interface	0.079 (0.020)	None
M1	Test stopped at 121.5 days		None	None	None
E1	Test stopped at 114.6 days		None	None	None
T1	0.9	0.9	2 Interfaces	Hairline	0.426 (1.06)
U1	13.8	11.7	1 Interface	0.020 (0.050)	None
U2	Test stopped at 120 days		None	None	None
C1	15.9	4.1	1 Interface	Hairline	0.001 (0.003)

*All initial cracking occurred at one or two interfaces. No additional locations were documented past the first crack location(s) for any material.

Visual cracking was observed in the G1 ring at 2.9 days after grout placement. Cracks appeared at both interfaces between the grout and the precast concrete. As shown in Figure 27 one strain gage observed a decrease in shrinkage strain approximately 1.6 days after casting. The strain in each of the three operational gages was approximately -8 microstrain indicating there was little strain in the ring. The other two gages continued measuring around -15 microstrain. This small amount of residual strain after cracking may have been due to a chemical bond between the grout and the adjacent steel ring. Note that the fourth strain gage failed at the start of the test.

Visual cracking was seen in the G2 ring at 2.0 days after grout placement. Cracks appeared at both interfaces between the grout and the precast concrete. As shown in Figure 28 there was a slight expansion up until the rings were demolded at day 1. From demolding until approximately 1.5 days, the strain reduced by 5 to 10 microstrain in three gages while remaining generally constant in the fourth gage. After cracking the strain in two gages began to increase slightly while the strain in the other two gages remained near zero.

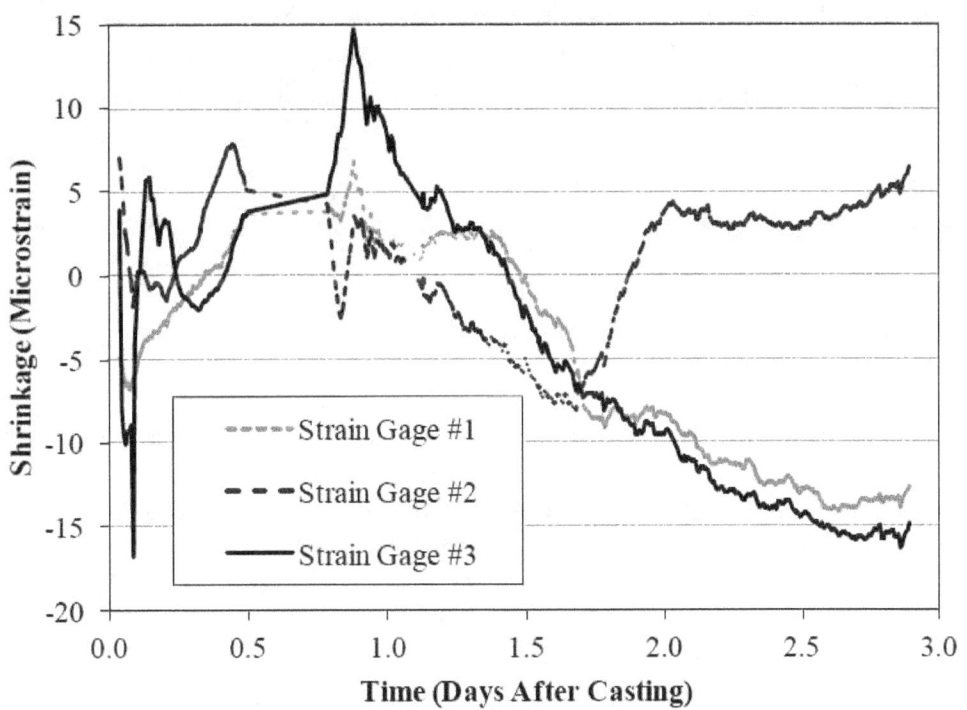

Figure 27. Graph. Restrained shrinkage bond results for G1.

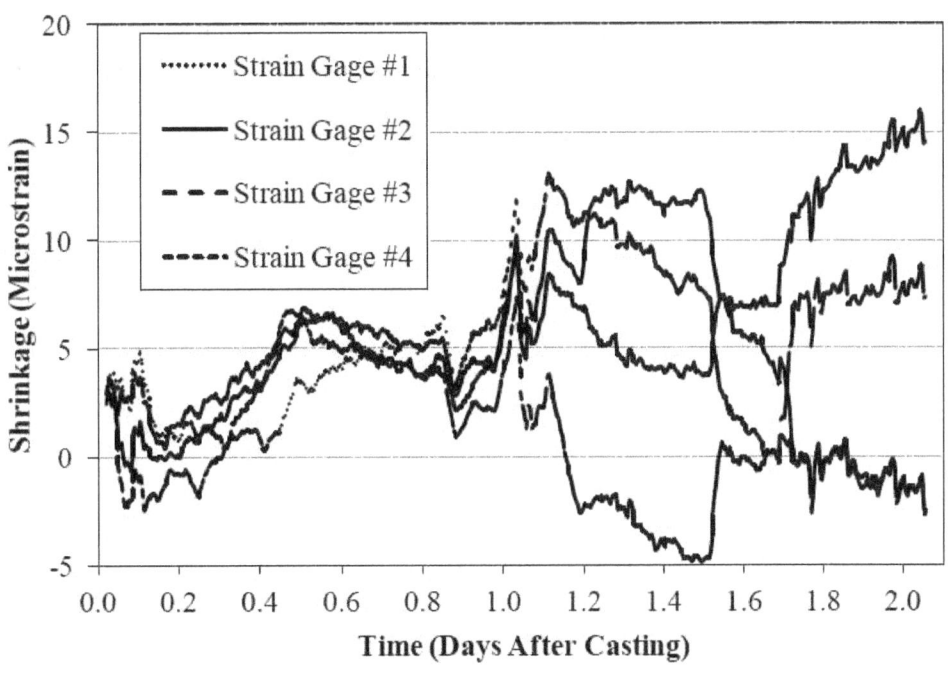

Figure 28. Graph. Restrained shrinkage bond results for G2.

The T1 ring did not return any usable strain gage data. The grout ring was visually cracked upon demolding at 24 hours. The strain gages never registered any major strain change indicating that cracking occurred very soon after the grout was placed. This follows the pattern recorded with ASTM C157-08 indicating T1 shrinks and cracks very early.

For two of the field-cast grout materials tested, namely M1 and E1, no cracking was visually observed throughout the duration of the test. The strain versus time plots for these materials showed more consistent patterns confirming that cracking did not occur. Based on the strain data for these two rings, it is proposed that this modified test method works better for rings that do not crack within two to three days.

As seen in Figure 29, the M1 bonded ring expanded approximately 20 microstrain in the first twenty days of testing. Beginning at approximately 28 days after placement, data was captured periodically. Strain from day 28 until day 120 show a very slow loss in strain with no significant discontinuities. One of the gages continues increasing in strain up to about 45 microstrain in expansion at 120 days. The other three gages range from 0 to 20 microstrain at 120 days. None of the strain gages indicated that M1 was exhibiting shrinkage. It must be noted that the implemented test method is not intended to measure expansive materials. However, the results do confirm that the material does not appear to be shrinking over time and that there was no cracking in the ring.

As shown in Figure 30, the E1 specimen exhibited shrinkage which generated between 5 and 70 microstrain in the steel ring during the first day after casting. One gage returned almost no change in strain after that, while the other three gages continued to record a slow rate of shrinkage. Approximately 35 days after casting, the strain gages ceased observing shrinkage and began observing either slight expansion or dimensionally stable responses. Following the first twenty-four hours there was never a significant discontinuity in strain in the rings. This confirms the observation that the ring did not crack. It also follows the pattern observed in the ASTM C157-08 test wherein this material does exhibit initial shrinkage which is followed by a dimensionally stable response.

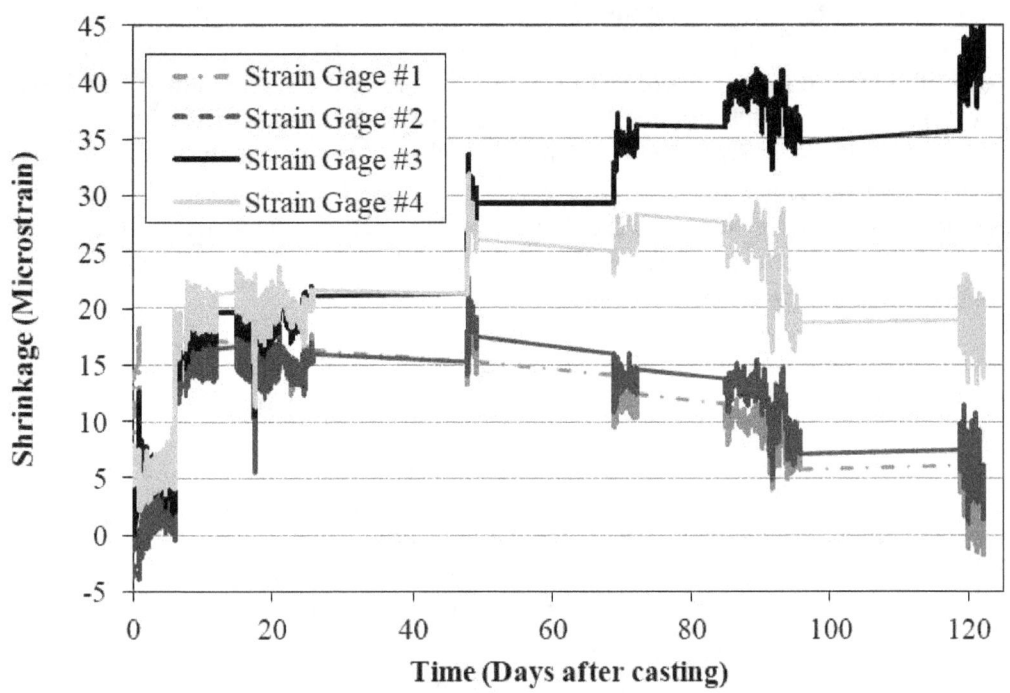

Figure 29. Graph. Restrained shrinkage bond results for M1.

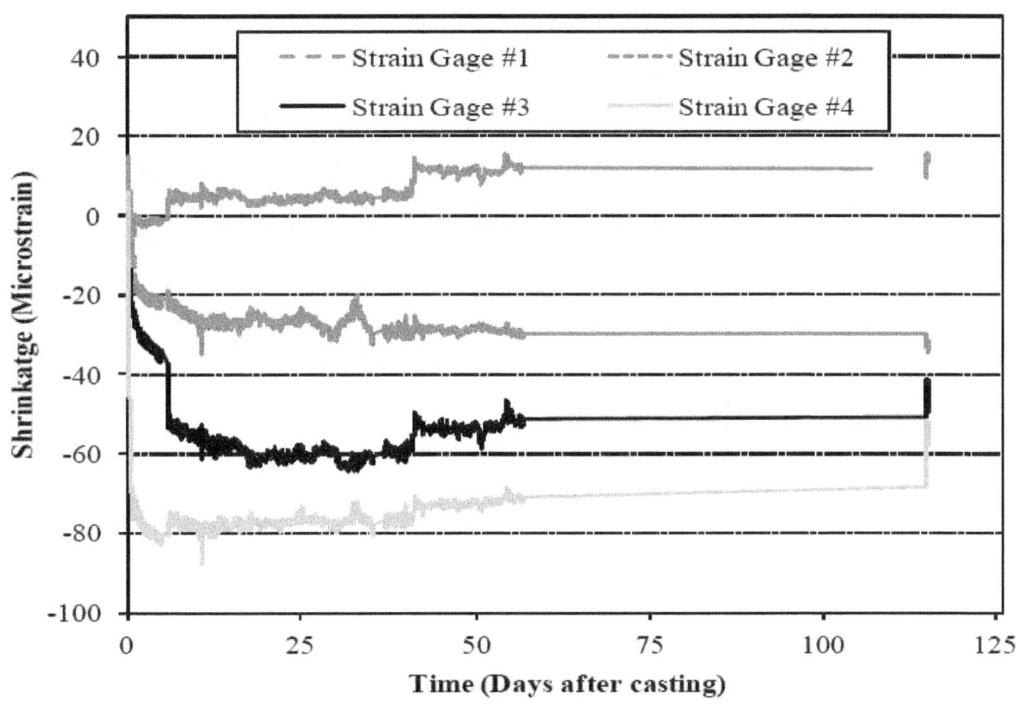

Figure 30. Graph. Restrained shrinkage bond results for E1.

Visual cracking was observed in the U1 test specimen 13.8 days after casting. A small crack appeared at one of the interfaces between U1 and the deck concrete. As shown in Figure 31, three of the four strain gages observed behavior consistent with restrained shrinkage cracking of the specimen at 11.7 days after casting. Specifically, these gages observed an increase in tension in the ring of between 5 and 10 microstrain, which is consistent with the loss of the compressive forces in the ring which had been generated by shrinkage of the specimen. After this, all four gages begin to observe an unloading of the steel ring. By 19 days after casting, three of the four gages exhibit readings close to zero and the strain on the fourth gage is significantly reduced. This data appears to confirm the visual cracking timeline.

No visual cracking was observed in the U2 test specimen through 120 days after casting. A problem with the electronic data collection system resulted in the strain data not being collected for this specimen.

Visual cracking was seen in the C1 test specimen 15.9 days after casting. A small crack appeared at one of the interfaces between the deck concrete and C1. As shown in Figure 32 the steel ring experienced a slight expansion until demolding. After demolding the gages do not follow any clear patterns. Two gages begin to observe shrinkage, one observed little change, and one begins to observe expansive strain. At 4.1 days after casting, the two gages that had observed shrinkage showed an expansive jump in strain. The results thereafter are less clear, with one strain gage continuing to observe shrinkage through 12.5 days after casting. After this, all gages exhibit movement toward zero strain indicating that the steel ring is observing reduced load levels. These results are inconclusive, but indicate that the cracking may have occurred as early at 4.1 days after casting.

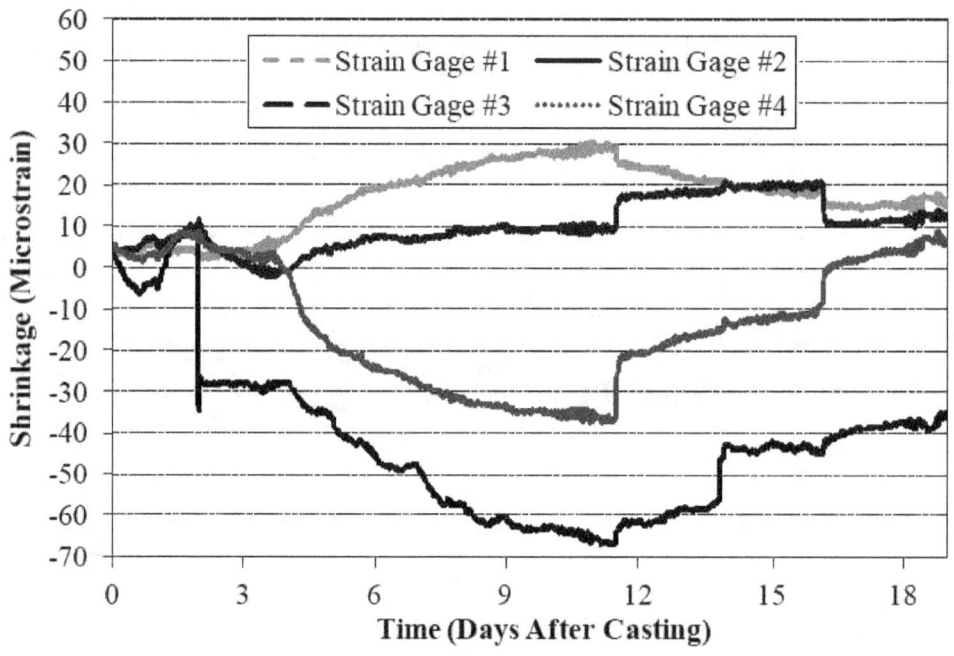

Figure 31. Graph. Restrained shrinkage bond results for U1.

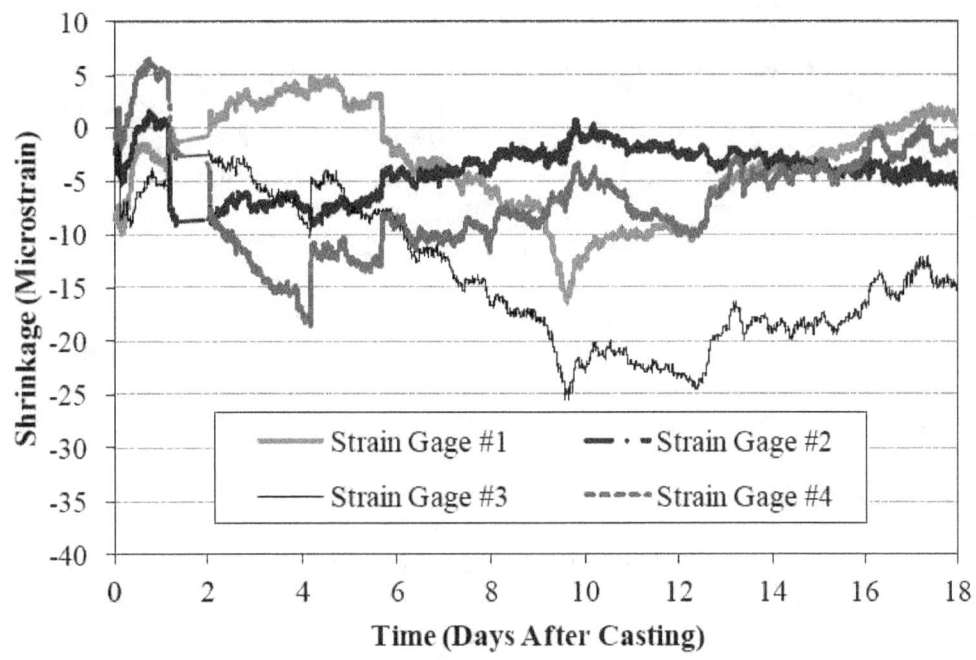

Figure 32. Graph. Restrained shrinkage bond results for C1.

The strain gages on the steel rings experienced strain changes of less than 25 microstrain throughout testing for G1, G2, and C1. They all indicated cracking within the first week. In each case, the strain gage data is difficult to conclusively interpret. However, with all three materials, the visual cracks were very clear and easy to identify. The use of a visual crack identification method is extremely important when engaging this modified test procedure to assess the performance of materials which exhibit limited bond and/or crack soon after placement.

In future testing, it may be prudent to demold the rings sooner than 24 hours for materials that set quickly. The data for the rings prior to demolding tended to be erratic because of the confinement caused by the outer ring. Some materials could be demolded within 6 to 12 hours providing a clearer image of strain change early in its life. The ASTM C1581-09a test method that this bond test was based upon is typically used for concrete materials that have longer set times. A slight adjustment in demolding may be wise given half of the material set within a few hours and then cracked within a few days after casting.

Another suggestion would be to not demold the half rings for the precast concrete prior to pouring the bonded materials. By demolding the precast concrete half specimen, the formwork had to be readjusted when re-forming the test setup. A method could be devised to sandblast or otherwise prepare the interface surface without removing the formwork. This may eliminate some of the erratic data that appeared from some of the strain gages.

DURABILITY TESTING

Freeze-Thaw Resistance

The freeze-thaw resistance was measured using ASTM C666-03 *Standard Test Method for Resistance of Concrete to Rapid Freezing and Thawing*. A representative set of four grouts, specifically grouts G1, E1, M1, and U2, were tested for freeze-thaw resistance. In this test method, the prismatic test specimens are subjected to freezing and thawing while submerged in a water bath. The aggressive environment created by this tests helps to assess whether the grout is capable of resisting the expansion effects that can occur when water within the pore structure of the material freezes. By periodically measuring the change in the resonant frequency of the prism, the test method provides an indication of the internal degradation that can occur in a specimen over the course of the test. The test method is normally run for 300 cycles of freezing and thawing; however, in this case the cycling was extended to 600 cycles for three of the grouts. This test extension was intended to provide a means of differentiating performance for the three grouts that performed well during the initial 300 cycles.

The prismatic specimens had nominal dimensions of 3 in. by 4 in. by 16 in. (76.2 x 101.6 x 406.4 mm). Three prisms were tested for each grout. Procedure A was followed wherein the test specimens are both frozen and thawed in water. Readings were collected on average every 12 cycles during the first 300 cycles of testing, and every 20 cycles during the final 300 cycles of testing. Both the dynamic modulus of elasticity and the mass of each specimen was collected at each cycle interval.

The relative dynamic modulus of elasticity results are presented in Figure 33, while the mass change results are presented in Figure 34. All three prisms from the U2, G1, and E1 sets reached the conclusion of the 600 freeze/thaw cycles and could have been subjected to continued testing. At 300 cycles, the relative dynamic modulus of elasticity values for the U2, G1, and E1 prisms were 101%, 99%, and 93%, respectively. At 600 cycles, the values were 99%, 97%, and 90%, respectively. The three prisms from the M1 set degraded rapidly, expressing twelve percent drop in relative dynamic modulus within six cycles, and a 75 percent decrease by 22 cycles.

The mass change results also provide an indication of the performance of the test specimens. The U2 and E1 specimens show very little change in mass throughout the testing. Combined with the visual observations for these specimens, this result is indicative of the fact that the specimens neither lost significant mass from the exterior of the specimen nor gained significant mass by absorbing water. The G1 specimens showed an initial slight increase in mass, followed by a continual slight decrease in mass throughout the conclusion of the testing. The M1 specimens exhibited a more rapid increase in mass until the tests on these prisms were stopped.

A photograph of the 4 in. by 16 in. (101.6 x 406.4 mm) side of one specimen in each set is provided in Figure 35 through Figure 38. The U2 and E1 prisms are observed to have sustained very little surface degradation. The G1 prism had begun to show some surface roughening by the conclusion of the testing. The M1 prism shown lasted the longest of the M1 set, undergoing 45 cycles. By the cessation of testing on this prism, it had lost a large portion of its mass and no longer resembled a prismatic element.

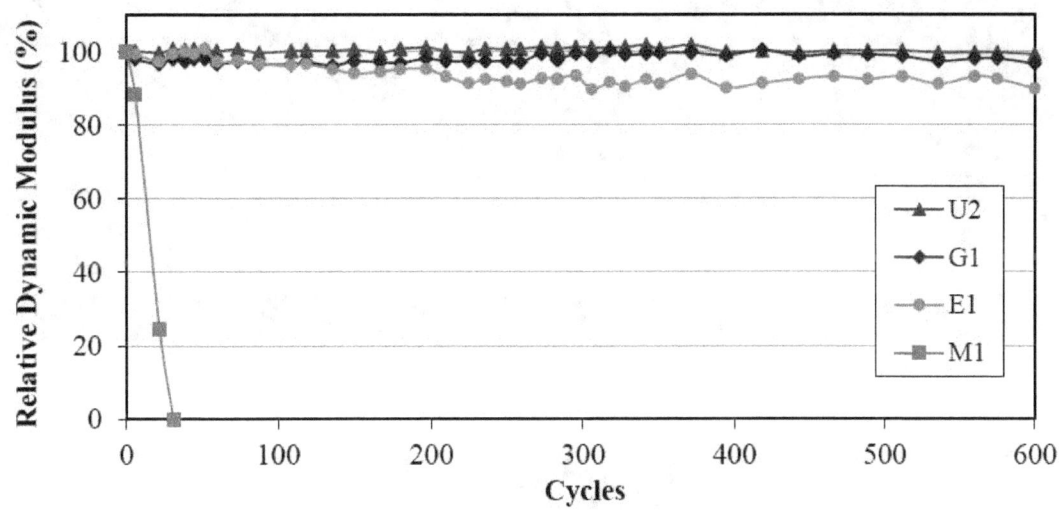

Figure 33. Graph. Relative dynamic modulus of elasticity of freeze/thaw prisms.

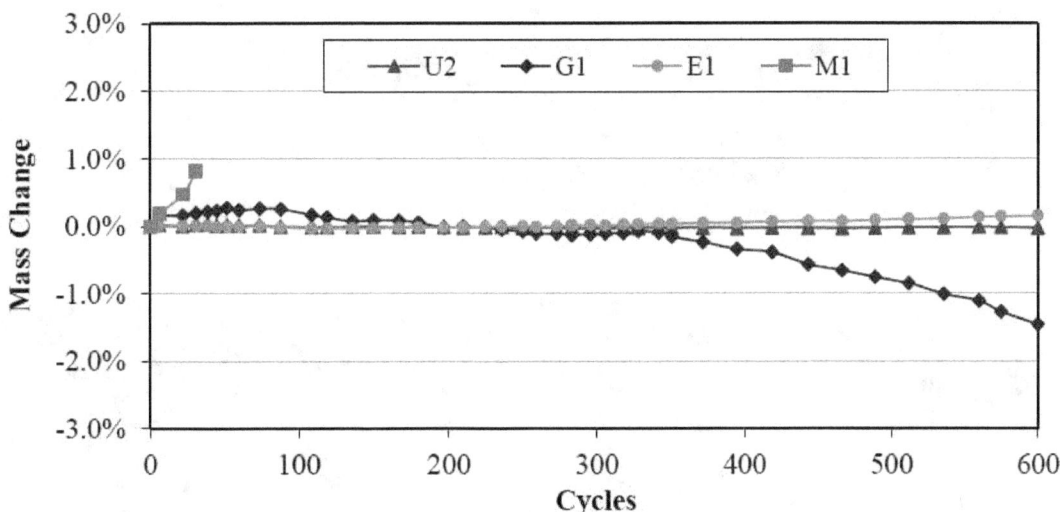

Figure 34. Graph. Mass change of freeze/thaw prisms.

Figure 35. Photo. U2 prism after the completion of 600 freeze/thaw cycles.

Figure 36. Photo. G1 prism after the completion of 600 freeze/thaw cycles.

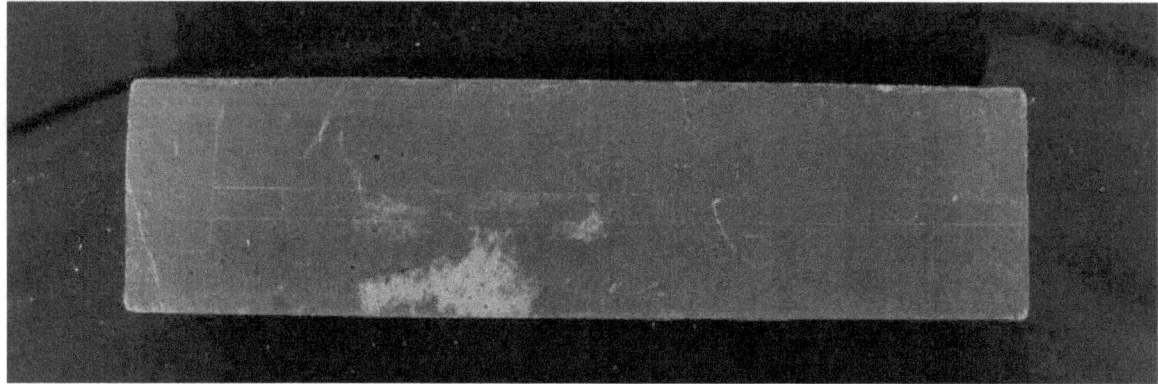

Figure 37. Photo. E1 prism after the completion of 600 freeze/thaw cycles.

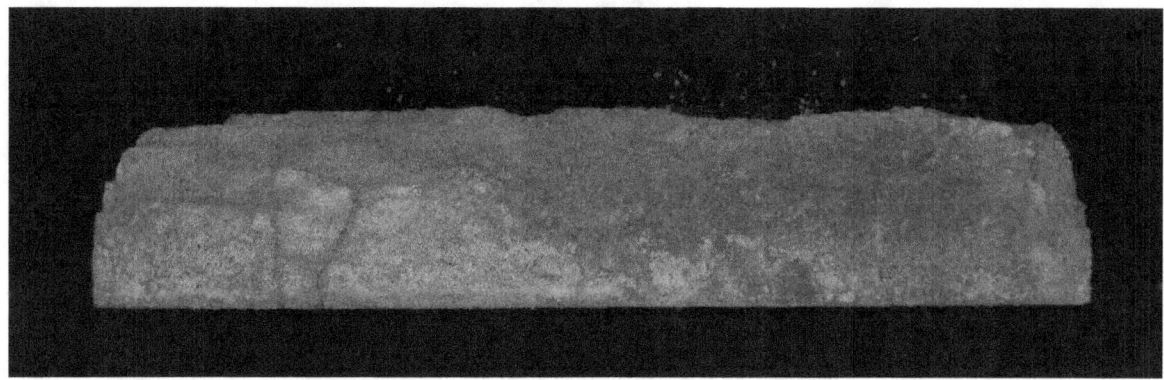

Figure 38. Photo. M1 prism after the completion of 45 freeze/thaw cycles.

Rapid Chloride Penetrability

The ability the field-cast grouts to resist the penetration of chloride ions was assessed through the ASTM C1202-10 *Standard Test Method for Electrical Indication of Concrete's Ability to Resist Chloride Ion Penetration*. The same representative set of grouts, specifically grouts G1, E1, M1, and U2, were subjected to this test method. This test approximates the resistance that a concrete may exhibit to chloride ion penetration by measuring the amount of electrical current that passes through a 51-mm (2-inch) thick slice of concrete over 6 hours. A 60-volt direct current (DC) potential is applied across the slice, while a sodium chloride solution is applied to one side of the slice, and a sodium hydroxide solution is applied to the other side.

These tests were completed on slices from 4-inch (102-mm) diameter cylinders that were cast alongside the freeze-thaw prisms previously discussed. After demolding, the cylinders were placed in a lime water bath until testing. Each 4-inch (102-mm) diameter cylinder was cast to have a length of 8 inches (203 mm). One slice was cut from the top, middle, and bottom of each cylinder according to the as-cast orientation of the cylinder. The three slices from one cylinder for each grout were tested at 57 days and at 240 days after casting. The top slice from one cylinder for each grout was also tested at 126 days after casting. Note that the charge passed results obtained through this test were corrected according to the test method to be representative of the charge passing through a 3.75-inch (95-mm) diameter slice.

The results of these tests are presented in Table 27. The E1 grout has a non-conductive epoxy-based matrix and thus conducted no appreciable charge throughout the duration of the test. The U2 grout passed a very low charge according to the test method. The M1 grout passed a low charge, and the G1 grout passed a high charge.

Table 27. Rapid Chloride Ion Penetrability Results.

Grout	Age (days)	Slice Location	Coulombs Passed	Chloride Ion Penetrability
U2	57	Top	189	Very Low
	57	Middle	807	Very Low
	57	Bottom	*	
	126	Top	526	Very Low
	240	Top	389	Very Low
	240	Middle	350	Very Low
	240	Bottom	301	Very Low
G1	57	Top	7306	High
	57	Middle	7083	High
	57	Bottom	8792	High
	126	Top	6996	High
	240	Top	4466	High
	240	Middle	4461	High
	240	Bottom	3114	Moderate
E1	57	Top	0	Negligible
	57	Middle	1	Negligible
	57	Bottom	0	Negligible
	126	Top	0	Negligible
	240	Top	1	Negligible
	240	Middle	0	Negligible
	240	Bottom	15	Negligible
M1	57	Top	1604	Low
	57	Middle	1756	Low
	57	Bottom	1528	Low
	126	Top	1091	Low
	240	Top	836	Very Low
	240	Middle	1075	Low
	240	Bottom	952	Very Low

* Data collection error resulted in no data being collected.

CHAPTER 4. RESULTS

The detailed results presented in Chapter 3 are compiled and presented concisely under relevant headings throughout this chapter. Topics include construction, material properties, bond strength, and durability.

CONSTRUCTION

Workability

All of the materials were workable and could be used in the field-casting of connections between precast concrete components. Every material except M1 remained workable for over 30 minutes. The results of the ASTM C1437 flow test are provided in Figure 39. These results, which include the full complement of 25 table drops, show that G1, G2, G3, T1, and U2 exhibited the maximum dynamic flow measurable via this test method. It must also be noted that T1 and U2 reached the full flow measurement without any drops of the table, indicating that they are much more similar to self-leveling materials than the conventional grouts which flowed less than 5.0 in. (12.7 cm) prior to the dropping of the table. Materials U1, M1, and E1 were stiffer, displaying an approximately 1 in (2.5 cm) increase in flow between the initial static measurement and the final dynamic measurement.

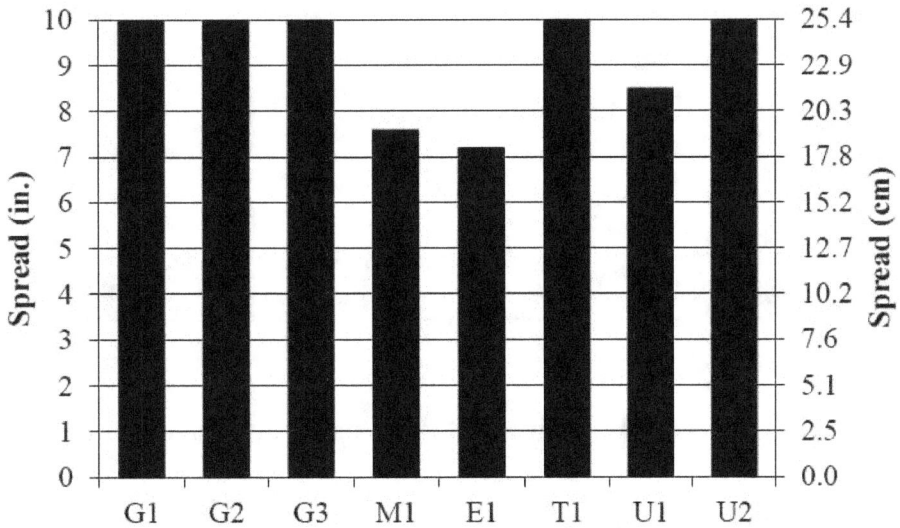

Figure 39. Graph. Spread measurements using ASTM C1437-07.

Recall that individual mix designs with specific water contents were engaged in this research effort. Many grouts, including materials G1, G2, G3, and T1 tested in this study, allow for a range of water contents. Different water contents result in different consistencies as well as differences in other material properties. Grouts consistencies can range from plastic to flowable to fluid, with higher water contents generally resulting in greater shrinkage and lower strength. Refer to the Appendix for manufacturer reported mix information and material properties relevant to different water contents for G1, G2, G3, and T1.

Cleanup

Materials G1, G2, G3, and C1 did not require any modified cleaning or casting procedures compared to typical concrete pours. U1 and U2 were not hard to clean but did contain steel fibers which required variations in cleaning and casting. T1 and E1 needed abrasion to clean from tools and formwork. M1 set very fast and required constant cleaning of tools to ensure future use. M1 and E1 were difficult to demold because they bonded well to the steel forms.

Set Time

The set times based on ASTM C403 demonstrated that grouts display a wide range of setting time behaviors. The M1 grout set within minutes, while the G2 and U1 materials did not reach initial set until more than 8 hours after mix initiation. Figure 40 provides a summary of the setting time results.

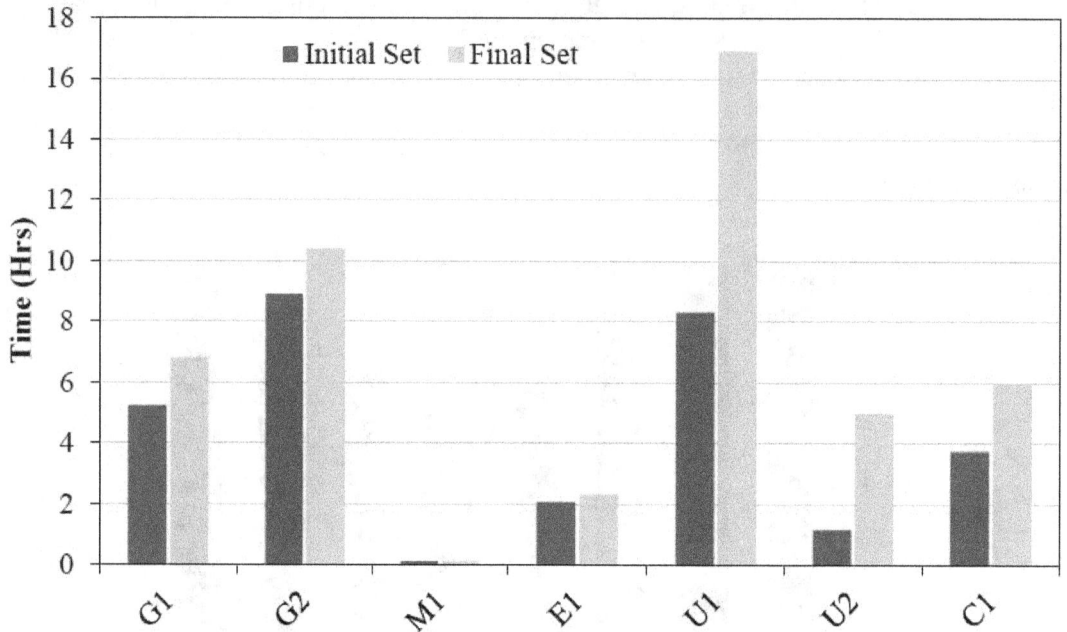

Figure 40. Graph. Set times based on ASTM C403-08.

Cost

The unit costs of the materials tested in this study at the time of acquisition in 2010 and 2011 in the Washington, D.C. metropolitan area are presented in Figure 41. All of the grouts cost significantly more than traditional ready-mix concrete. In general, the grouts fall in the 1000 to 2000 \$/yd3 range (1300 to 1600 \$/m^3), with the E1 grout being the outlier at nearly 4600 \$/yd^3 (6000 \$/m^3). Note that these material costs did not include transportation costs, handling costs, or taxes.

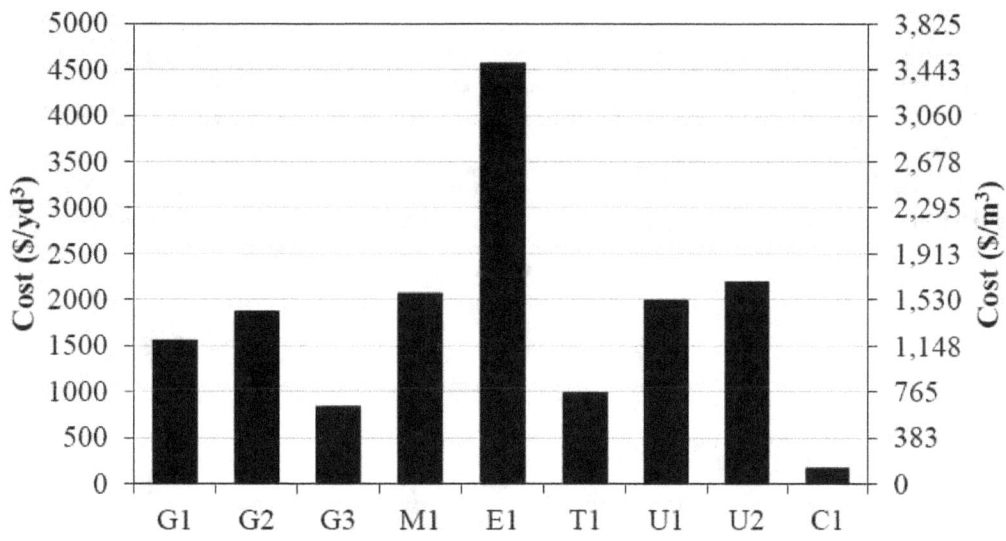

Figure 41. Graph. Price comparison of the materials.

MATERIAL PROPERTIES

Unit Weight
Aside from the UHPC materials, the grout materials studied herein tend to be between 10-30% lighter than a typical bridge deck concrete. The results are shown in Figure 42. The UHPC materials are slightly heavier than typical concrete.

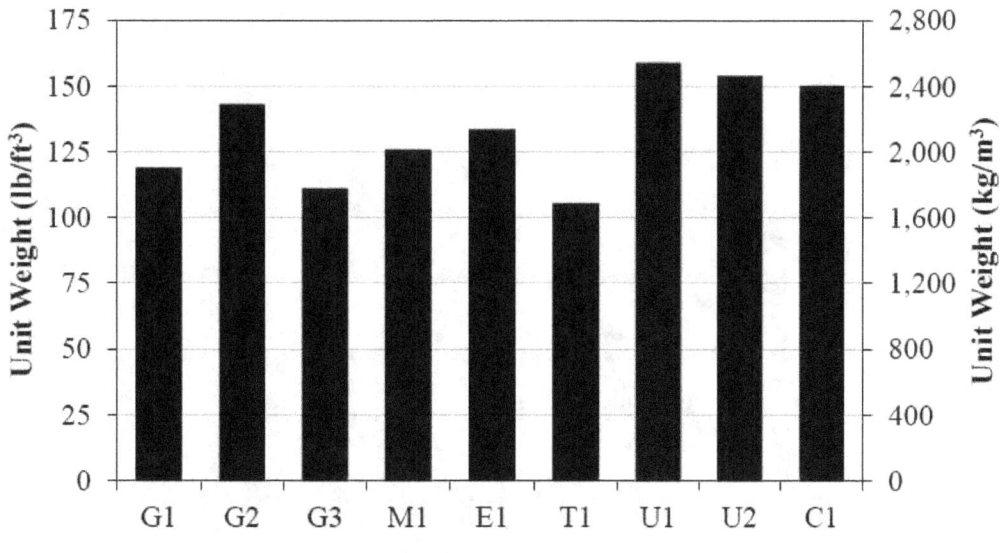

Figure 42. Graph. Unit weights.

Compressive Strength

The compressive strength results are presented in Figure 43. All of the materials tested had compressive strengths of at least 4 ksi (27.6 MPa) within 7 days. Material C1 had the lowest 7 day strength while E1 and U1 exhibited compressive strengths over 14 ksi (97 MPa).

M1, E1, and U2 had over 8 ksi (55.2 MPa) compressive strength within 24 hours of mix initiation. Materials G1, G2, and G3 had at least 3 ksi (20.7 MPa) of compressive strength at 24 hours.

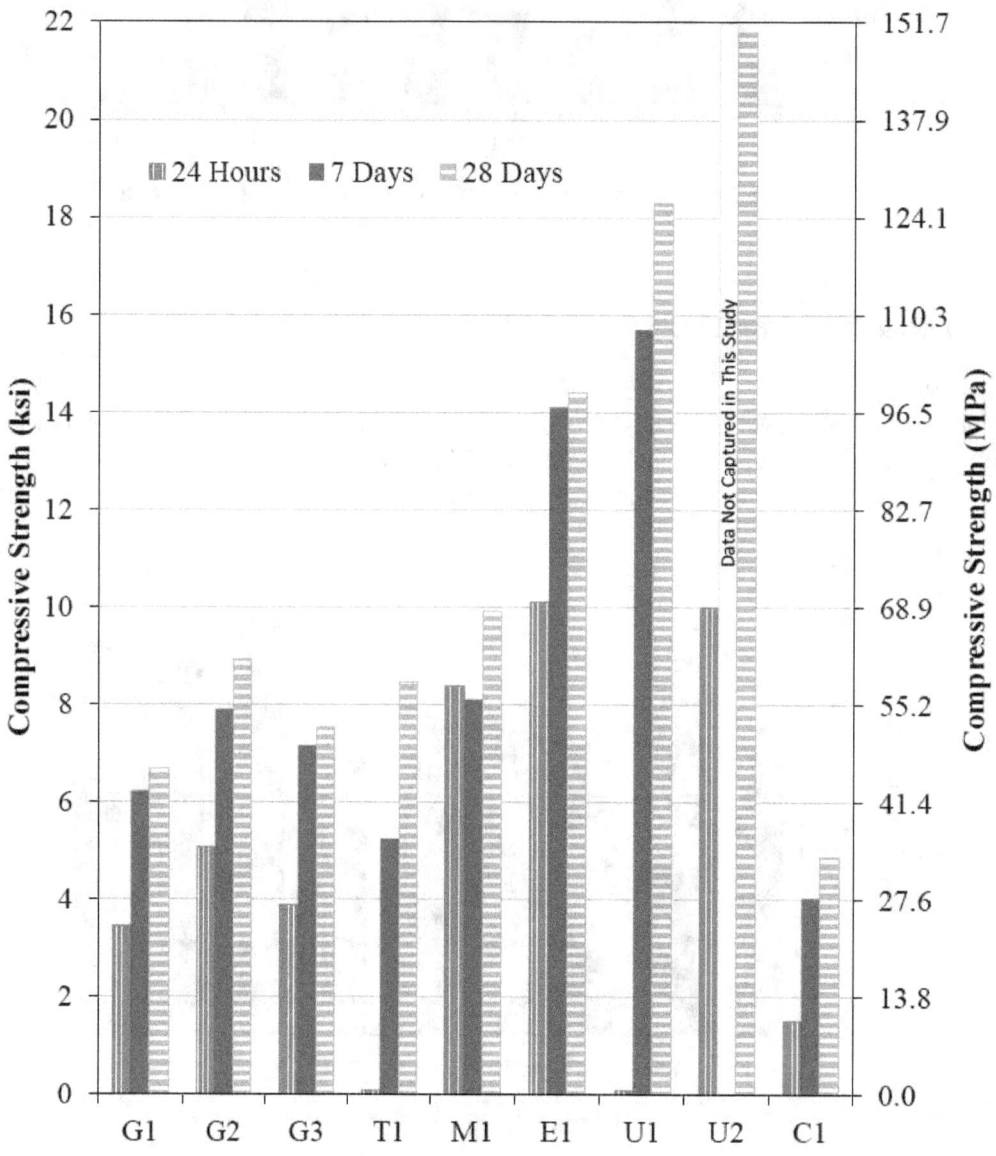

Figure 43. Graph. Compressive strength results.

It must be noted that the rate of compressive strength gain is highly dependent on the curing conditions to which the grout material is subjected. The chemical reactions inherent in these types of grout materials are temperature dependent and thus will be accelerated by warmer temperatures and delayed by colder temperatures.

Tensile Strength
The summary of the ASTM C496 splitting tensile strength results are presented in Figure 44. Note that this test method must be modified to be appropriate for fiber reinforced concretes, and thus results for U1 and U2 are not reported. Material E1 exhibited approximately 2000 psi (13.8 MPa) of splitting tensile strength at both 24 hours and 28 days after mix initiation. The other grouts all exhibited 24 hour tensile strengths between 330 and 435 psi (2.28 to 3.00 MPa) and 475 and 665 psi (3.28 and 4.59 MPa) at 28 days.

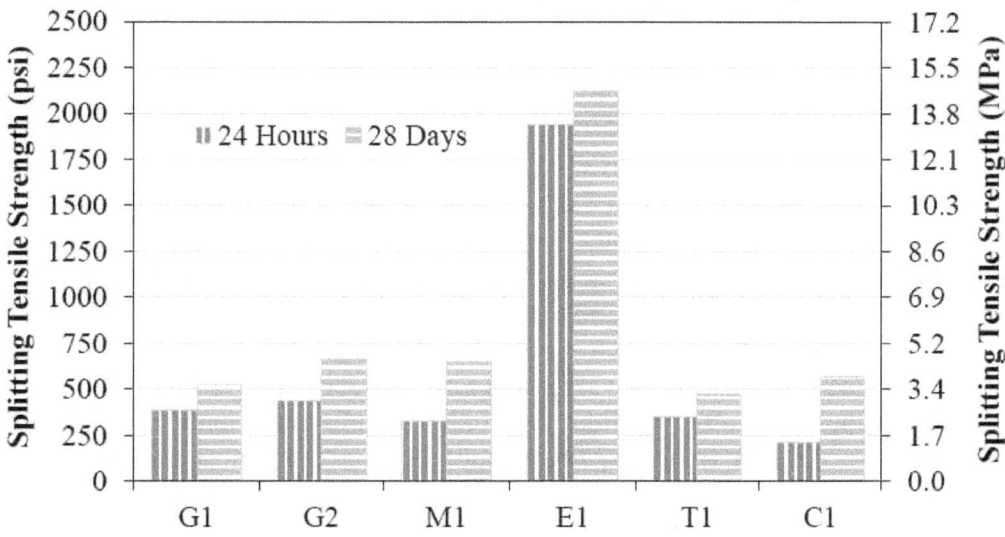

Figure 44. Graph. Splitting tensile strength results.

Modulus of Elasticity
The 28-day modulus of elasticity test results are presented in Figure 45. Materials U1 and U2 exhibit a high modulus of elasticity commensurate with their high compressive strengths. Materials M1 and E1 express stiffness values similar to that normally expected from conventional concrete. The conventional grout materials, which contain no coarse aggregate and are thus effectively mortars, exhibit reduced modulus values commensurate with the level of stiffness that is expected from mortars.

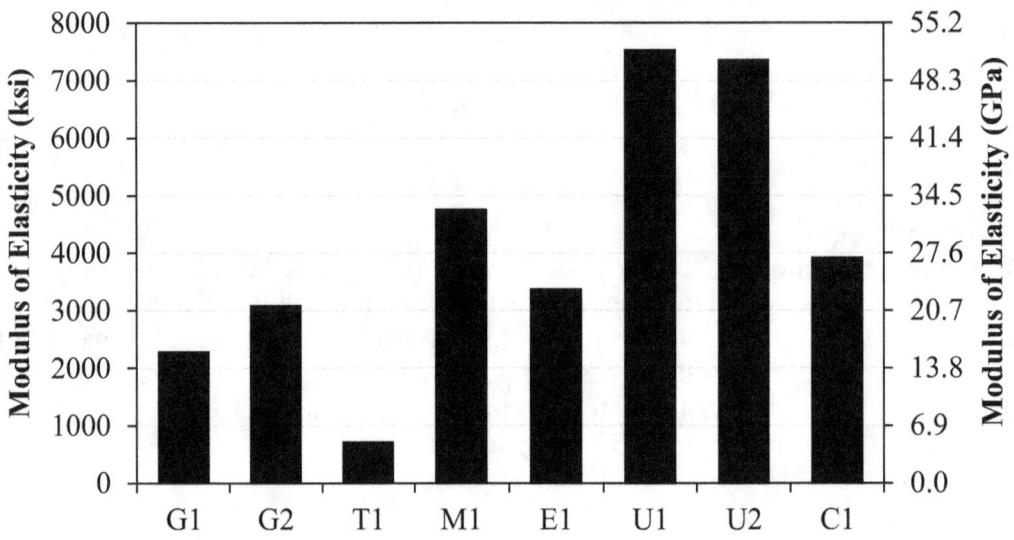

Figure 45. Graph. Modulus of elasticity results from 28 day tests.

Shrinkage

The majority of the volume change in M1 or E1 occurred within the first 24 hours. As such, their results as captured via the ASTM C157-08 shrinkage test were close to zero because shrinkage behavior during the initial 24 hours is excluded from this test method. The ASTM C157-08 test method is generally more applicable to deck concretes like C1 and most conventional grouts like G1, G2, and G3, as these cementitious materials tend to exhibit comparatively decreased rates of strength gain and shrinkage in the first 24 hours after casting. T1 had extremely high shrinkage values and cracked considerably. Figure 46 provides a summary of the results observed from the strain measurements captured via the vibrating wire gages in the modified ASTM C157-08 method. Material M1 had a net increase while C1 and E1 had values of approximately 400 microstrain. The expansion of M1 may be greater than what was measured because the test setup was not designed for expansive materials. Material T1 had a strain value greater than the limit of the gages of 4000 microstrain.

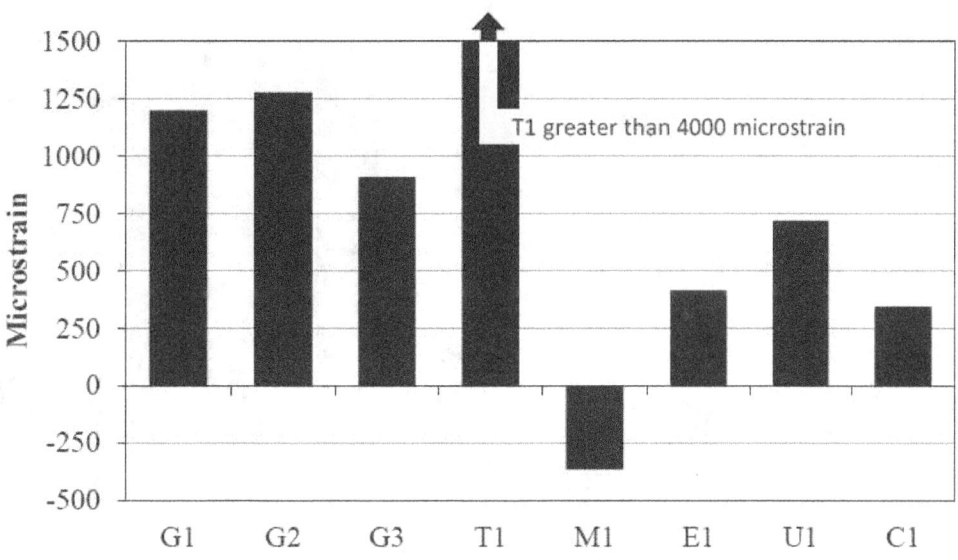

Figure 46. Graph. Strain 28 days after casting using the modified ASTM C157 with vibrating wire gages.

BOND STRENGTH

Slant Cylinder Compression Test

The summary of 28-day results from the slant cylinder bond strength test is presented in Figure 47. This test provides an indication of the bond strength under compressive and shearing loads up to the limiting strength of the precast concrete to which the secondary material bonded. The precast surface was sandblasted prior to casting of the secondary material. Specimens in the C1, T1, G1, and G2 sets exhibited cracking at the bond interface prior to the start of the test. They, along with the G3 and M1, all failed along the bonded interface. M1 had a significantly reduced bond due to specimen fabrication errors caused by its limited working time and thus the result is not plotted here. E1 and U1 had sufficient bond strength to result in failures through the precast concrete.

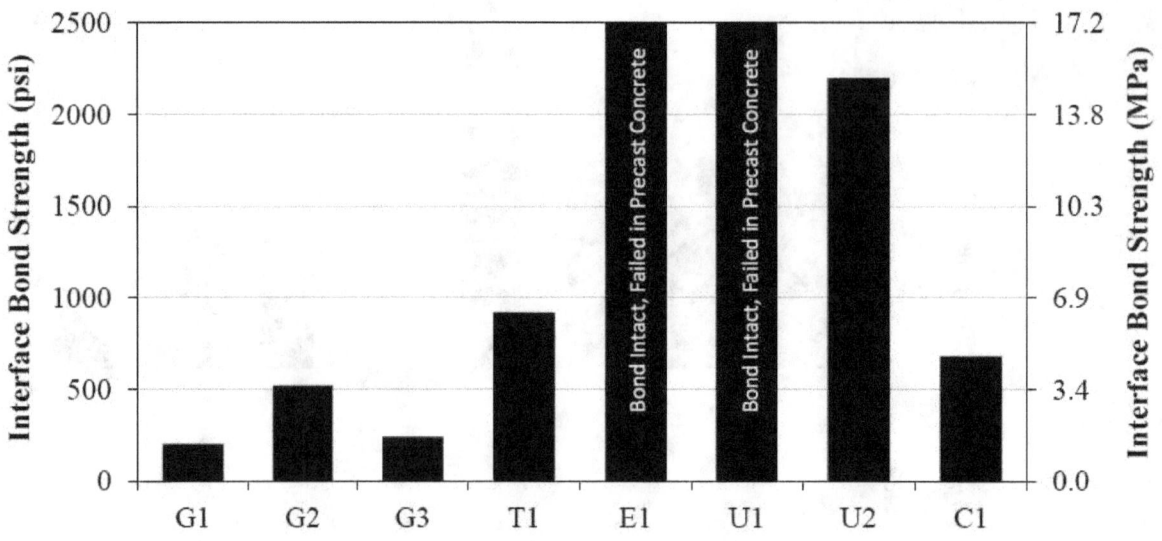

Figure 47. Graph. Slant cylinder bond strength based on ASTM C882.

Splitting Tensile Bond Test

The splitting cylinder tensile bond strength test results are presented in Figure 48. This test provides an indication of the tensile bond strength of the secondary cast material to the precast concrete as a result of the biaxial state of stress generated in the cylindrical specimen due to the transverse loading. The bonding surfaces of the precast half-cylinders were sandblasted prior to the secondary cast. The tests were completed 28-days after the casting of the secondary material. The splitting tensile bond failure plane was along the interface for G1, G2, G3, T1, M1, and C1. For specimens in the G1, G2, G3, T1, and M1 groups, small cracks were observed along the interface prior to the start of the tests. E1, U1, and U2 failed within the paste of the precast concrete paste matrix adjacent to the bonded interface. This indicates that their bond strength was greater than the tensile strength of the precast concrete.

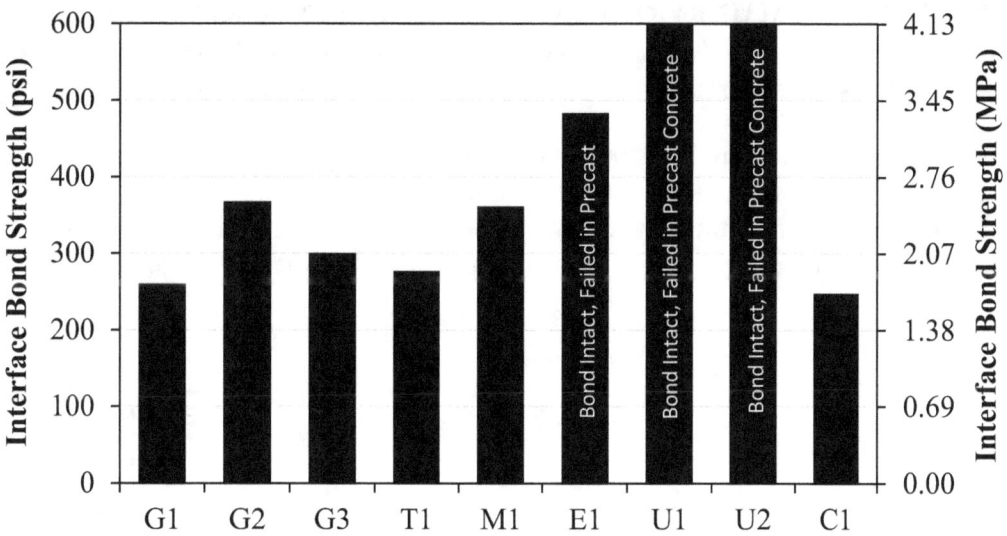

Figure 48. Graph. Splitting tensile bond strength based on ASTM C496.

DURABILITY

Limited durability testing was completed on a subset of the materials engaged in this overall research effort. Specifically, grouts G1, M1, E1, and U2 were subjected to ASTM C666 freeze/thaw resistance testing and to ASTM C1202 chloride ion penetrability testing. A range of performances were observed in these tests.

In the freeze/thaw testing, the G1, E1, and U2 materials reached 600 cycles while retaining at least 90% of their initial relative dynamic modulus. Conversely, M1 was quickly damaged by the freezing and thawing, with the three tested prisms all shedding material, cracking, and failing within 50 cycles.

In the chloride ion penetrability testing, the four grouts each exhibited different resistance levels. E1, with its non-conductive, non-porous matrix, did not conduct electrical current and thus was observed to exhibit negligible chloride ion penetrability. U2, with its dense cementitious matrix, achieved a very low penetrability. M1 generally exhibited a low penetrability with between 800 and 1800 Coulombs passed. The conventional grout, G1, was observed to be far more susceptible to chloride ion penetrability with over 7000 Coulombs passes during 57-day tests. The conductivity did decrease by the 240-day tests where only approximately 4000 Coulombs were passed.

GRAPHICAL SUMMARY OF GROUT PERFORMANCE RESULTS

Throughout this research effort, the performance of a variety of field-castable grout-type material has been investigated through the conduct of various material characterization tests.

Figure 49 presents a graphical summary of the results of this research effort. The results for each parameter denoted on the left are plotted on the adjacent linear scale. The limits of the linear scale are shown. This graphic allows for visual interpretation of the overall results of the research program, thus providing for a simplified grasp of the performances of each grout.

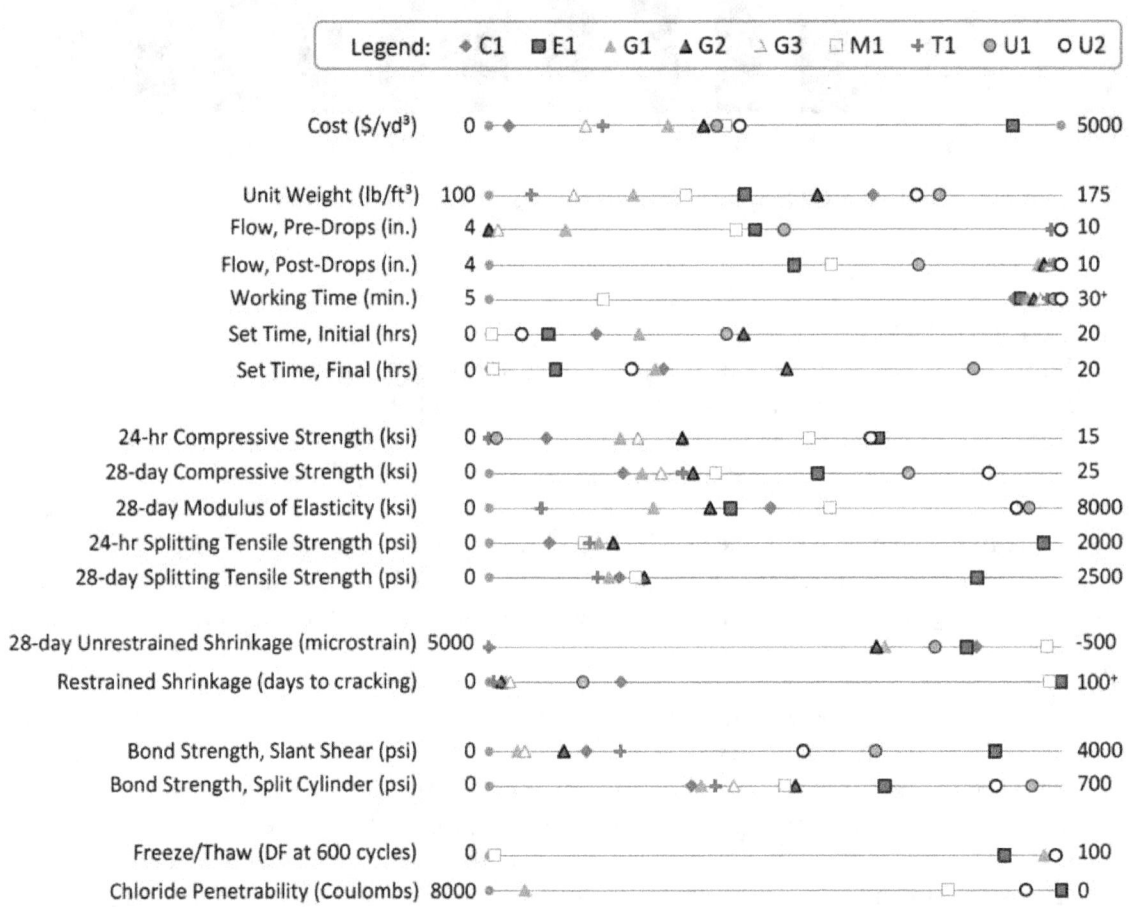

Figure 49. Graph. Graphical representation of the performance of the tested materials.

CHAPTER 5. CONCLUSIONS

SUMMARY

The test program discussed herein focused on characterizing basic mechanical, dimensional stability, and bond properties for eight field-cast grouts that could be used in connecting precast concrete bridge components. The results demonstrate that the material characteristics, practical construction considerations, and cost can vary widely. These results and others must be carefully considered when selecting the appropriate grout to use in a particular construction project.

For accelerated construction projects requiring high compressive strengths within one day, E1 and U2 displayed acceptable properties. E1 had sufficient strength gain, was one of the most dimensionally stable materials, had good workability, and high tensile strength. The material also developed strong bonds with the precast concrete in all three bond tests and expressed good durability properties. However, its comparatively high cost may limit its viability.

U2 also displayed appropriate strength gain, was comparatively dimensionally stable, had good workability, high tensile strength, and high modulus of elasticity. The material contains internal fiber reinforcement that can arrest any cracking that may occur. This material expressed good durability properties. The material developed strong bonds to the precast concrete and had a unit price approximately half that of E1.

An alternative for this type of project and for other projects requiring exceptionally rapid strength gain is M1. The greatest concerns with this material relate to constructability considerations, including its very limited working time, and to its durability. The limited work time created problems when trying to quickly cast the material in the formwork. The freeze/thaw test specimens rapidly deteriorated.

For construction that allows a longer cure time, U1 is a viable choice. This material has high compressive strength, high tensile strength, and internal fiber reinforcement that can arrest any cracking that may occur. The bond strengths of U1 were high in the slant cylinder bond and split cylinder bond tests. Total shrinkage, although less than observed with the conventional grouts G1 and G2, is greater than that exhibited by M1 and E1.

C1 mix performed as well as the standard grouts in most cases. The conventional grouts shrank more, had only modestly higher compressive strengths and bond strengths, cracked earlier, and cost substantially more. However, the rheological properties of conventional concretes combined with the sizes of aggregated commonly included in conventional concrete, present fundamental hurdles that are addressed by the conventional grouts.

RECOMMENDATION

Owners, specifiers, and designers considering the use of field-cast grouts for PBES connections should carefully consider the performance measures that are of greatest interest before, during, and after deployment of the application. Many classes of grout-type materials are available, with each offering different performance levels relative to different performance metrics. In all cases, it is important to ensure that the connection design is constructible, durable, and economical in near term and over the life of the constructed facility.

ONGOING AND FUTURE RESEARCH
The research discussed herein is part of a larger program aimed at facilitating the use of prefabricated bridge elements and systems. Other portions of the program are further investigating the interface bond performance and the shrinkage performance of grouts. Future phases of the program will investigate structural performance of various field-cast connection details both as subcomponents and as part of full-scale bridge structures.

ACKNOWLEDGEMENT
The research project discussed herein could not have been completed were it not for the dedicated support of the federal and contract staff associated with the FHWA Structural Concrete Research Program. Special recognition goes to Dr. Matthew Swenty and Brenton Stone, each of whom oversaw significant aspects of the research discussed herein. Additional engineering assistance was provided by Dr. Gary Greene and Dr. Jussara Tanesi. Technical assistance with the casting, preparation and testing of specimens was provided by Bradford Tschetter, Daniel Balcha, Tim Tuggle, Brian Story, Kevin Deasy, and Paul Ryberg.

The publication of this report does not necessarily indicate approval or endorsement of the findings, opinions, conclusions, or recommendations either inferred or specifically expressed herein by the Federal Highway Administration or the United States Government.

REFERENCES

BS 4449 Steel Bars for the Reinforcement of Concrete. (1978). London: British Standards Institution.

Achillides, Z., & Pilakoutas, K. (2004). Bond Behavior of Fiber Reinforced Polymer Bars Under Direct Pullout Conditions. *Journal of Composites for Construction*, 173-181.

ACI Committee 408. (2003). *Bond and Development of Straight Reinforcing Bars in Tension.* Farmington Hills: American Concrete Institute.

American Concrete Institute (ACI). (2005). *Manual of Concrete Practice* (Vols. ACI 345.2R-98). Detroit: MI.

Arnold, D. J. (1980). *Concrete Bridge Decks: Does Structural Vibration Plus Excess Water Form the Fracture Plane.* Michigan Department of Transportation.

ASTM_International. (1991, June). Standard Test Method for Comparing Concretes on the Basis of the Bond Developed with Reinforcing Steel. *ASTM C 234*. West Conshohocken, PA, US: ASTM International.

Badie, S. S., Tadros, M. K., & Baishya, M. C. (1998). NUDECK- An Efficient and Economical Precast Prestressed Bridge Deck System. *PCI Journal, 43*(5), 56-83.

Biswas, M. (1986). Precast Bridge Deck Design Systems. *Journal of the Prestressed Concrete Institute, 31*(2), 40-94.

Cairns, J., & Abdullah, R. (1995, June 6). An Evaluation of Bond Pullout Tests and Their Relevance to Structural Performance. *Structural Engineer, 73*(11), 179-185.

Carter, J. W., Pilgrim, T., Hubbard, F. K., Poehnelt, T., & Oliva, M. (2007, January-February). Wisconsin's Use of Full-Depth Precast Concrete Deck Panels Keeps Interstate 90 Open to Traffic. *PCI Journal*, 2-16.

Csagoly, P. F., Campbell, T. I., & Agarwal, A. C. (1972). *Bridge Vibration Study.* Downsview: Ontario Ministry of Transportation and Communications.

Culmo, M. P. (2009). *Connection Details for Prefabricated Bridge Elements and Systems.* Mclean: US Department of Transportation, Federal Highway Administration.

Deaver, R. (1982). *Brige Widening Study.* Georgia Department of Transportation.

DeMurphy, M. L., Kim, J., Sang, Z., & Xiao, C. (2010). *Determining More Effective Approaches and Materials for Grouting Shear Keys.* Harrisburg: Pennsylvania Transportation Institute.

Devore, J. L. (1995). *Probability and Statistics for Engineering and the Sciences.* Belmont: Wadsworth Publishing Co.

Feldman, L. R., & Bartlett, F. M. (2005, Nov./Dec.). Bond Strength Variability in Pullout Specimens with Plain Reinforcement. *ACI Structural Journal*, 860-867.

Ferguson, P. M., Turpin, R. D., & Thompson, J. N. (1954). Minimum Bar Spacing as a Function of Bond and Shear Strength. *Journal of the American Concrete Institute*, 869-887.

FHWA. (2010, November 4). *Every Day Counts.* (Federal Highway Administration) Retrieved November 19, 2010, from https://www.fhwa.dot.gov/everydaycounts/

Furr, H. L., & Fouad, H. F. (1981). *Bridge Slab Concrete Placed Adjacent to Moving Live Loads.* Texas Transportation Institute.

Geissert, D. G., Li, S. E., Frantz, G. C., & Stephens, J. E. (1999). Splitting Prism Test Method to Evaluate Concrete-to-Concrete Bond Strength. *ACI Materials Journal*, 359-366.

Graybeal, B. (2006). Practical Means for Determination of the Tensile Behavior of Ultra-High Performance Concrete. *3*(8), 10 pp.

Graybeal, B. (2012). *Construction of Field-Cast Ultra-High Performance Concrete Connections.* Federal Highway Administration, HRT-12-038, 8 pp.

Graybeal, B., and Stone, B. (2012). *Compression Response of a Rapid-Strengthening Ultra-High Performance Concrete Formulation.* Federal Highway Admininstration, National Technical Information Service Accession No. PB2012-112545, 66 pp.

Gulyas, R. J., Wirthlin, G. J., & Champa, J. T. (1995). Evaluation of Keyway Grout Test Methods for Precast Concrete Bridges. *PCI Journal*, 44-57.

Hamad, B. S., & Sabbah, S. M. (1998). Bond of Reinforcement in Eccentric Pullout Silica Fume Concrete Specimens. *Materials and Structures*, 707-713.

Hao, Q., Wang, Y., He, Z., & Ou, J. (2009). Bond Strength of Glass Fiber Reinforced Polymer Ribbed Rebars in Normal Strength Concrete. *Construction and Building Materials*, 865-871.

Harsh, S., & Darwin, D. (1983). *Effects of Innovative Construction Procedures on Concrete Bridge Decks, Final Report Part II, Effects of Traffic Induced Vibrations on Bridge Deck Repairs.* Lawrence: University of Kansas Center for Research.

Harsh, S., & Darwin, D. (1984). *Effects of Traffic Induced Vibrations on Bridge Deck Repairs.* Lawrence: University of Kansas Center for Research.

Issa, M. A. (1999, May). Investigation of Cracking in Concrete Bridge Decks at Early Age. *Journal of Bridge Engineering, 4*(2), 116-124.

Issa, M. A., Ralph, A., Thomas, D., Shaker, A., & Islam, M. S. (2007, May/June). Full-Scale Testing of Prefabricated Full-Depth Precast Concrete Bridge Deck Panel System. *ACI Structural Journal*, 324-332.

Issa, M. A., Yousif, A. A., Issa, M. A., Kaspar, I. I., & Khayyat, S. Y. (1995). Field Performance of Full Depth Precast Concrete Panels in Bridge Deck Reconstruction. *PCI Journal, 40*(3), 82-108.

Kwan, A. K., & Ng, P. L. (2006). Reducing Damage to Concrete Stitches in Bridge Decks. *Bridge Engineering*(BE2), 53-62.

Lachemi, M., Bae, S., & Hossain, K. M. (2009). Stee-Concrete Bond Strength of Lightweight Self-Consolidating Concrete. *Materials and Structures*, 1015-1023.

Larnach, W. J. (1952, July). Change in Bond Strength Caused by Re-vibration of Concrete and the Vibration of Reinforcement. *Magazine of Concrete Research*, 17-21.

Lutz, L. A. (1970, November). Information on the Bond of Deformed Bars from Special Pullout Tests. *ACI Journal*, 885-887.

Ma, J. (2010). Durability Performance Criteria of Closure Pour Materials for CIP Connections. *PCI Bridge Conference.* Washington, D.C.: PCI.

Manning, D. G. (1981). *Effects of Traffic-Induced Vibrations on Bridge-Deck Repairs.* NCHRP.

Markowski, S. M. (2005). *Experimental and Analytical Study of Full-Depth Precast / Prestressed Concrete Deck Panels for Highway Bridges.* Madison: University of Wisconsin - Madison.

McMahon, J. E., & Womack, J. C. (1965). *Bridge Widening Problems, .* California Division of Highways.

Mindess, S., Young, F. J., & Darwin, D. (2003). *Concrete* (2 ed.). Upper Saddle River, NJ: Pearson Education, Inc.

Momayez, A., Ehsani, M. R., Ramezanianpour, A. A., & Rajaie, H. (2005). Comparisons of Methods for Evaluating Bond Strength Between Concrete Substrate and Repair Materials. *Cement and Concrete Research*, 748-757.

Montero, A. C. (1980). *Effect of Maintaining Traffic During Widening of Bridge Decks (A Case Study)*. Columbus: Ohio State University.

Prenger, H. B. (1992). *Bridge Deck Cracking*. Maryland Department of Transportation.

RILEM/CEB/FIP. (1978). Bond Test for Reinforcing Steel 2: Pullout Test. *Recommendation RC6*. Bagneux, France: RILEM/CEB/FIP.

Scholz, D. P., Wallenfelsz, J. A., Ligeron, & Roberts-Wollmann, C. L. (2007). *Recommendations for the Connection Between Full-Depth Precast Bridge Deck Panel Systems and Precast I-Beams*. Charlottesville: Virginia Transportation Research Council.

Silwerbrand, J. (1992). Influence of Traffic Vibrations on Repaired Concrete Bridge Decks. *Third International Workshop on Bridge Rehabilitation*, (pp. 416-474). Darmstadt, Germany.

Sullivan, S. (2007). *Construction and Behavior of Precast Bride Deck Panel Systems*. Blacksburg, VA, VA: Virginia Tech.

Swenty, M. K. (2009). *The Investigation of Transverse Joints and Grouts on Full Depth Concrete Bridge Deck Panels*. Blacksburg: Virginia Tech.

Swenty, M., and Graybeal, B. (2012). *Influence of Differential Deflection on Staged Construction Deck-Level Connections*. Federal Highway Administration, National Technical Information Service Accession No. PB2012-111528, 82 pp.

Tholen, M. L., & Darwin, D. (1996). *Effects of Deformation Properties on the Bond of Reinforcing Bars*. Lawrence, KS: University of Kansas Center for Research, Inc.

Whiffen, A. C., & Leonard, D. R. (1971). *A Survey of Traffic Induced Vibrations*. England: Transport and Road Research Laboratory.

Wight, J. K., & MacGregor, J. G. (2009). *Reinforced Concrete*. Upper Saddle River: Pearson Prentice Hall.

APPENDIX A

A.1 FIVE STAR GROUT MANUFACTURER'S DATA SHEET

FIVE STAR® GROUT
High Performance Precision Nonshrink Grout

PRODUCT DESCRIPTION

Five Star® Grout is the industry's leading cement-based, nonmetallic, nonshrink grout for supporting machinery and equipment. It is formulated with Air Release technology that combines high performance with the greatest reliability. When tested in accordance with ASTM C 827, Five Star Grout exhibits positive expansion. Five Star Grout meets the performance requirements of ASTM C 1107-02 Grades A, B and C, ASTM C 1107-07, and CRD-C 621-93 specifications for nonshrink grout over a wide temperature range, 40°F - 90°F (4°C - 32°C).

ADVANTAGES

- Air release technology per ACI 351.1 R
- 95% Effective Bearing Area (EBA) is typically achieved following proper grouting procedures
- Provides placement versatility: pour, pump or dry pack
- 45 minute working time
- Permanent support for machinery requiring precision alignment
- Does not contain gas generating additives, such as aluminum powder
- Nonshrink from the time of placement

USES

- Grouting of machinery and equipment to maintain precision alignment
- Nonshrink grouting of structural steel and precast concrete
- Grouting of anchors and dowels
- Support of tanks and vessels

PACKAGING AND YIELD

Five Star Grout is packaged in heavy-duty, polyethylene lined bags and is available in 50 lb (22.7 kg) units yielding approximately 0.5 cubic feet (14.1 liters), or 100 lb (45.4 kg) units yielding approximately 1.0 cubic foot (28.3 liters) of hardened material at maximum water content.

SHELF LIFE

One year in original unopened packaging when stored in dry conditions; high relative humidity will reduce shelf life.

TYPICAL PROPERTIES AT 70°F (21°C)		
Early Height Change, ASTM C 827	0.0 to 4.0%	
Hardened Height Change, ASTM C 1090	0.0 to 0.3%	
Effective Bearing Area	95%	
Bond Strength, ASTM C 882, 28 Days	2000 psi (13.8 MPa)	
Pull-out Strength, Shear Bond with #5 deformed bar, 7 Days	2400 psi (16.6 MPa)	
Compressive Strength, ASTM C 942 (C109 Restrained)	Minimum Water psi (MPa)	Maximum Water psi (MPa)
1 Day	4000 (27.6)	2500 (17.3)
3 Days	5500 (38.0)	3500 (24.1)
7 Days	6500 (44.9)	5000 (34.5)
28 Days	8000 (55.2)	6500 (44.9)
Working Time at 70°F (21°C)	45 minutes	

The data shown above reflects typical results based on laboratory testing under controlled conditions. Reasonable variations from the data shown may result. Test methods are modified where applicable.

PLACEMENT GUIDELINES

1. **SURFACE PREPARATION:** All surfaces in contact with Five Star® Grout shall be free of oil, grease, laitance and other contaminants. Concrete must be clean, sound and roughened to ensure a good bond. Soak concrete surfaces for 8 to 24 hours prior to application with liberal quantities of potable water, leaving the concrete saturated and free of standing water.

2. **FORMWORK:** Formwork shall be constructed of rigid non-absorbent materials, securely anchored, liquid-tight and strong enough to resist forces developed during grout placement. The clearance between formwork and baseplate shall be sufficient to allow for a headbox. The clearance for remaining sides shall be one to two inches (25 - 50 mm). Areas where bond is not desired must be treated with form oil, paste wax or similar material. Isolation joints may be necessary depending on pour dimensions. Contact the Five Star Engineering and Technical Center for further information.

3. **MIXING:** Mix Five Star Grout thoroughly for approximately four to five minutes to a uniform consistency with a mortar mixer (stationary barrel with moving blades). A drill and paddle mixer is acceptable for single bag mixes. For optimum performance, maintain grout at ambient temperatures between 40°F and 90°F (4°C and 32°C). Use heated or chilled water to help adjust working time. Mix Five Star Grout with 7 - 11 quarts potable water per 100 lb. bag (3 1/2 to 5 1/2 quarts per 50 lb. bag). Do not exceed maximum recommended amount of mixing water as stated on the package or add an amount that will cause segregation. Working time is approximately 45 minutes at 70°F (21°C). Follow printed instructions on the package. Always add mixing water first to mixer followed by grout.

4. **PLACEMENT PROCEDURES:** Five Star Grout may be dry packed, poured or pumped into place. Minimum placement thickness for Five Star Grout is 1 inch (25 mm). For pours over 6 inches (150 mm) in depth Five Star Grout should be extended with a clean, damp coarse aggregate meeting the requirements of ASTM C 33. Refer to the Five Star Technical Bulletin "Cement Grout Aggregate Extension" for guidelines.

5. **POST-PLACEMENT PROCEDURES:** Five Star Grout shall be wet cured for a minimum of three days, or coated with an approved curing compound meeting the requirements of ASTM C 309 after a minimum 24 hour wet cure. In-service operation may begin immediately after the required grout strength has been reached.

NOTE: PRIOR TO APPLICATION, READ ALL PRODUCT PACKAGING THOROUGHLY. For more detailed placement procedures, refer to Design-A-Spec™ installation guidelines or call the Five Star Engineering and Technical Service Center at (800) 243-2206.

CONSIDERATIONS

- If temperatures of equipment and surfaces are not between 40°F and 90°F (4°C and 32°C) at time of placement, refer to Design-A-Spec™ for cold and hot weather grouting procedures, or call the Five Star Engineering and Technical Service Center at (800) 243-2206.
- Never exceed the maximum water content as stated on the bag or add an amount that will cause segregation. Construction practices dictate concrete foundation should achieve its design strength before grouting.

CAUTION

Contains cementitious material and crystalline silica. International Agency for Research on Cancer has determined that there is sufficient evidence for the carcinogenicity of inhaled crystalline silica to humans. Take appropriate measures to avoid breathing dust. Avoid contact with eyes and contact with skin. In case of contact with eyes, immediately flush with plenty of water for at least 15 minutes. Immediately call a physician. Wash skin thoroughly after handling. Keep product out of reach of children. **PRIOR TO USE, REFER TO MATERIAL SAFETY DATA SHEET.**

For worldwide availability, additional product information and technical support, contact your local Five Star distributor, local sales representative, or you may call Five Star's Engineering and Technical Service Center at (800) 243-2206.

WARRANTY: "FIVE STAR PRODUCTS INC. (FSP) PRODUCTS ARE MANUFACTURED TO BE FREE OF MANUFACTURING DEFECTS AND TO MEET FSP'S CURRENT PUBLISHED PHYSICAL PROPERTIES WHEN APPLIED IN ACCORDANCE WITH FSP'S DIRECTIONS AND TESTED IN ACCORDANCE WITH ASTM AND FSP STANDARDS. HOWEVER, SHOULD THERE BE DEFECTS OF MANUFACTURING OF ANY KIND, THE SOLE RIGHT OF THE USER WILL BE TO RETURN ALL MATERIALS ALLEGED TO BE DEFECTIVE, FREIGHT PREPAID TO FSP, FOR REPLACEMENT. THERE ARE NO OTHER WARRANTIES BY FSP OF ANY NATURE WHATSOEVER, EXPRESS OR IMPLIED, INCLUDING ANY WARRANTY OF MERCHANTABILITY OR FITNESS FOR A PARTICULAR PURPOSE IN CONNECTION WITH THIS PRODUCT. FSP SHALL NOT BE LIABLE FOR DAMAGES OF ANY SORT, INCLUDING PUNITIVE, ACTUAL, REMOTE, OR CONSEQUENTIAL DAMAGES, RESULTING FROM ANY CLAIMS OF BREACH OF CONTRACT, BREACH OF ANY WARRANTY, WHETHER EXPRESSED OR IMPLIED, INCLUDING ANY WARRANTY OF MERCHANTABILITY OR FITNESS FOR A PARTICULAR PURPOSE OR FROM ANY OTHER CAUSE WHATSOEVER. FSP SHALL ALSO NOT BE RESPONSIBLE FOR USE OF THIS PRODUCT IN A MANNER TO INFRINGE ON ANY PATENT HELD BY OTHERS."

Five Star Products, Inc.
Corporate Headquarters
750 Commerce Drive
Fairfield, CT 06825 USA
Tel: 203-336-7900 · Fax: 203-336-7930
http://www.fivestarproducts.com

©2009 Five Star Products, Inc. (07/01/09)
American Owned & Operated

A.2 EMBECO 885 GROUT MANUFATURER'S DATA SHEET

The Chemical Company

PRODUCT DATA

3 03 62 16 Metallic Non-Shrink Grouting

EMBECO® 885

High-precision, nonshrink metallic-aggregate grout with extended working time

Description

Embeco® 885 is a cement-based metallic-aggregate grout with an extended working time. It is ideally suited for grouting machines or plates requiring optimum toughness and precision load-bearing support, including machine bases subject to thermal movement. Embeco® 885 grout meets the requirements of ASTM C 1107 and the U.S. Army Corp of Engineers CRD C 621, Grades B and C.

Yield

One 55 lb (25 kg) bag of Embeco® 885 grout mixed with approximately 10 lbs (4.5 kg) or 1.2 gallons (4.5 L) of water yields approximately 0.43 ft^2 (0.012 m^3) of grout.

One 3,300 lb super sack yields approximately 1 cubic yard (0.72 m^3).

Packaging

55 lb (25 kg) multi-wall paper bags
3,300 lb (1,500 kg) bulk bags

Shelf Life

1 year when properly stored

Storage

Store in unopened bags in clean, dry conditions.

Features

- High fluidity
- Extended 30 minute working time
- High tolerance for wetting and drying cycles
- Hardens free of bleeding, segregation, or settlement shrinkage
- High tolerance to thermal movement, effects of heating and cooling
- High-quality well-graded blend of metallic-and-quartz aggregate
- Sulfate resistant

Benefits

- Ease of placement; self-consolidating
- Ensures proper placement under a variety of conditions
- Tolerates wet environments
- Provides maximum effective bearing area for optimum load transfer
- Ideal for harsh manufacturing environments
- Provides high strength, impact resistance; handles dynamic and repetitive loads
- For use in marine, wastewater, and sulfate-containing soil environments

Where to Use

INDUSTRIES

- Power generation
- Pulp and paper mills
- Steel and cement mills
- Stamping and machining
- Water and waste treatment
- Metal recycling
- General construction

APPLICATION

- Where high strength and impact resistance are required
- Where a nonshrink grout is needed to achieve maximum bearing for optimum load transfer
- Applications requiring a pumpable metallic grout with extended working time
- Compressors and generators
- Pump bases and drive motors
- Coal pulverizers
- Tank bases
- Conveyors
- Grouting anchor bolts, rebar and dowel rods

LOCATION

- Interior or exterior

How to Apply

Surface Preparation

1. Steel must be free of dirt, oil, grease, or other contaminants. Substrate must be fully cured (28 days).

2. The surface to be grouted must be clean, SSD, strong, and roughened to a CSP of 5 – 9 following ICRI Guideline 03732 to permit proper bond. For freshly placed concrete, consider using Liquid Surface Etchant (see Form No. 1020198) to achieve the required surface profile.

3. When dynamic, shear, or tensile forces are anticipated, concrete surfaces should be chipped with a "chisel-point" hammer to a roughness of (plus or minus) 3/8" (10 mm). Verify the absence of bruising according to ICRI Guideline 03732.

4. Concrete surfaces should be rough and saturated (ponded) with clean water for 24 hours just before grouting.

5. All freestanding water must be removed from the foundation and bolt holes immediately before grouting.

6. Anchor bolt holes must be grouted and sufficiently set before the major portion of the grout is placed.

Technical Data

Composition

Embeco® 885 is a hydraulic cement-based metallic-aggregate grout.

Compliances

- CRD C 621, Grades B and C
- ASTM C 1107, Grades B and C
- City of Los Angeles Research Report Number RR 23137

Test Data

PROPERTY	RESULTS			TEST METHODS
Compressive strengths, psi (MPa)				ASTM C 942, according to ASTM C 1107
		Consistency		
	Plastic[1]	Flowable[2]	Fluid[3]	
1 day	5,000 (34)	5,000 (34)	4,000 (28)	
3 days	7,000 (48)	6,000 (41)	5,000 (34)	
7 days	9,000 (62)	8,000 (55)	7,000 (48)	
28 days	11,000 (76)	10,000 (69)	9,000 (62)	
Volume change				ASTM C 1090
	% Change	% Requirement of ASTM C 1107		
1 day	> 0	0.0 – 0.30		
3 days	0.05	0.0 – 0.30		
14 days	0.07	0.0 – 0.30		
28 days	0.08	0.0 – 0.30		
Setting time, hr:min				ASTM C 191
		Consistency		
	Plastic[1]	Flowable[2]	Fluid[3]	
Initial set	3:30	5:00	5:30	
Final set	4:30	6:00	7:00	
Flexural strength,* psi (MPa)				ASTM C 78
3 days	880 (6.1)			
7 days	1,050 (7.2)			
28 days	1,150 (7.9)			
Modulus of elasticity,* psi (MPa)				ASTM C 469, modified
3 days	3.16×10^6 (2.18×10^4)			
7 days	3.50×10^6 (2.41×10^4)			
28 days	3.69×10^6 (2.54×10^4)			
Coefficient of thermal expansion,* in/in/° F (cm/cm/° C)	6.5×10^{-6} (11.7×10^{-6})			ASTM C 531
Punching shear strength,* psi (MPa). 3 by 3 by 11" (76 by 76 by 279 mm) beam				BASF Method
3 days	1,600 (11.0)			
7 days	1,800 (12.4)			
28 days	2,600 (17.9)			
Splitting tensile and tensile strength,* psi (MPa)				ASTM C 496 (splitting tensile) ASTM C 190 (tensile)
	Splitting Tensile	Tensile		
3 days	350 (2.4)	300 (2.1)		
7 days	490 (3.4)	400 (2.8)		
28 days	520 (3.6)	500 (3.4)		

[1] 100 – 125% flow on flow table per ASTM C 230
[2] 125 – 145% flow on flow table per ASTM C 230
[3] 25 to 30 seconds through flow cone per ASTM C 939
*Test conducted at a fluid consistency

This data was developed under controlled laboratory conditions. Expect reasonable variations

Test Data, continued

PROPERTY	RESULTS			TEST METHODS
Ultimate tensile strength and bond stress				ASTM E 488 Tests*
	Diameter, Inches	Depth, Inches	Tensile strength Lbs	Bond stress Psi
	5/8	4	29,200	3,718
	3/4	5	33,200	2,815
	1	7	58,500	2,660

* Average of 5 tests in ≥ 4,000 psi (27.6 MPa) concrete, using 125 ksi threaded rod in 2" diameter, damp, core-drilled holes.

Notes

1. Grout was mixed to a fluid consistency.

2. Recommended design stress: 1,750 psi.

3. Refer to the "Adhesive and Grouted Fastener Capacity Design Guidelines" for more detailed information.

4. Tensile tests with headed fasteners were governed by concrete failure.

Jobsite Testing

If strength tests must be made at the jobsite, use 2" (51 mm) metal cube molds as specified by ASTM C 942 or ASTM C 1107. DO NOT use cylinder molds. Control testing on the basis of the desired placing consistency rather than strictly on the water content.

---

7. Shade the foundation from sunlight 24 hours before and 24 hours after grouting.

Forming

1. Forms should be liquid tight and nonabsorbent. Seal forms with putty, sealant, caulk, polyurethane foam.

2. Moderately sized equipment should utilize a head form sloped at 45 degrees to enhance the grout placement. A moveable head box may provide additional head at minimum cost.

3. Side and end forms should be a minimum 1" (25 mm) distant horizontally from the object grouted to permit expulsion of air and any remaining saturation water as the grout is placed.

4. Leave a minimum of 2" between the bearing plate and the form to allow for ease of placement.

5. A minimum of 1" (51 mm) clearance is required where the grout will be placed.

6. Use sufficient bracing to prevent the grout from leaking or the form from moving.

7. Eliminate large, nonsupported grout areas wherever possible.

8. Extend forms a minimum of 1" (25 mm) higher than the bottom of the equipment being grouted.

9. Expansion joints may be necessary for both indoor and outdoor installation. Consult your local BASF field representative for suggestions and recommendations.

Temperature

1. For precision grouting, store and mix grout to produce the desired mixed-grout temperature. If bagged material is hot, use cold water. If bagged material is cold, use warm water. This will help achieve a mixed-product temperature as close to 70° F (21° C) as possible.

Recommended Temperature Guidelines for Precision Grouting

	MINIMUM °F (°C)	PREFERRED °F (°C)	MAXIMUM °F (°C)
Foundation and plates	45 (7)	50 – 80 (10 – 27)	90 (32)
Mixing water	45 (7)	50 – 80 (10 – 27)	90 (32)
Grout at mixed and placed temp.	45 (7)	50 – 90 (10 – 32)	90 (32)

2. If temperature extremes are anticipated or if special placement procedures are planned, contact your local BASF representative for assistance.

3. When grouting at minimum temperatures, take care to see that foundation, plate, and grout temperatures do not fall below 45° F (7° C) until after final set. Protect the grout from freezing (32° F or 0° C) until it has attained a compressive strength of 3,000 psi (21 MPa) in accordance with ASTM C 942 or C 1107.

Mixing

1. Place estimated water into the mixer (use potable water only), then slowly add the dry grout while mixing. For a fluid consistency, start with 9.2 lbs (4 kg) or 1.1 gallons (4.2 L) per 55 lb bag.

2. Water demand depends on mixing efficiency and material and ambient temperature conditions. Adjust the water to achieve the desired flow. Recommended flow is 25 – 30 seconds using the ASTM C 939 Flow-Cone Method. Use the minimum amount of water required to achieve the necessary placement consistency. Before placing grout below 45° F (7° C) and above 90° F (32° C), consult your BASF representative.

3. Moderate size batches of grout are best mixed in one or more clean mortar mixers. Large batches of grout are effectively, economically, and most efficiently mixed in ready-mix trucks using 3,300 lb (1,500 kg) bulk bags.

4. Mix grout a minimum of 5 minutes after all material and water are in mixer. Use mechanical mixer only.

5. Do not mix more grout than can be placed in approximately 30 minutes.

6. Transport by wheelbarrow or buckets, or pump to the equipment to be grouted. Minimize the transporting distance.

7. Do not retemper grout by adding water.

MBT PROTECTION & REPAIR PRODUCT DATA
EMBECO® 885

Application

1. Always place grout from only one side of the equipment to prevent entrapment of air or water beneath the equipment. Place Embeco® 885 grout in a continuous pour. Discard grout that becomes unworkable.

2. Immediately after placement, trim the surfaces with a trowel and cover the exposed grout with clean wet rags (not burlap). Maintain this moisture for 5 – 6 hours.

3. The grout should offer stiff resistance to penetration with a pointed mason's trowel before the grout forms are removed or excessive grout is cut back.

4. To further minimize the potential moisture loss within the grout, cure all exposed grout with an approved membrane curing compound (compliant with ASTM C 309 or preferably ASTM C 1315) immediately after the wet rags are removed.

5. Do not vibrate grout. Steel straps inserted under the plate may be used to aid in movement of the grout.

6. Consult your BASF representative before placing more than 6" (152 mm) in depth per lift.

For Best Performance

- For guidelines on specific anchor-bolt applications, contact BASF Technical Service.
- Do not add plasticizers, accelerators, retarders, or other additives unless advised in writing by BASF Technical Service.
- The water requirement may vary with mixing efficiency, temperature, and other variables.
- Hold a pre-job conference with your local representative to plan the installation. Hold conferences as early as possible before the installation of equipment, sole plates, or rail mounts. Conferences are important for applying the recommendations in this product data sheet to a given project, and they help ensure a placement of highest quality and lowest cost.
- The ambient and initial material temperature of the grout should be in the range of 45 to 90° F (7 to 32° C) for both mixing and placing. Ideally, use the amount of mixing water that is necessary to achieve a 25 – 30 second flow specified by ASTM C 939 (CRD C 611). For placement outside of 45 to 90° F (7 to 32° C), contact your local BASF representative.
- For pours greater than 6" (152 mm) deep, consult your local BASF representative for special precautions and installation procedures.
- When the grout will be in contact with steel stressed over 80,000 psi (550 MPa), use Masterflow® 816 cable grout or Masterflow® 1205, or Masterflow® 1341 post-tensioning cable grouts.
- Embeco® 885 is not intended for use as a floor topping or in large areas with exposed shoulders around baseplates. Where grout is exposed for shoulders, occasional hairline cracks may occur. Cracks may also occur near sharp corners of the baseplate and at anchor bolts. These superficial cracks are usually caused by temperature and moisture changes that affect the grout at exposed shoulders at a faster rate than the grout beneath the baseplate. They do not affect the structural, nonshrink, or vertical support provided by the grout if the foundation-preparation, placing, and curing procedures are properly carried out.
- Minimum placement depth is 1" (25 mm).
- Surfaces may discolor in certain environments; it is not an indication of product performance.
- Make certain the most current versions of product data sheet and MSDS are being used; call Customer Service (1-800-433-9517) to verify the most current version.
- Proper application is the responsibility of the user. Field visits by BASF personnel are for the purpose of making technical recommendations only and not for supervising or providing quality control on the jobsite.

Health and Safety

EMBECO® 885

WARNING

Embeco® 885 contains silica, crystalline quartz, portland cement; limestone; iron oxide; calcium oxide; gypsum; silica, amorphous; magnesium oxide.

Risks

Eye irritant. Skin irritant. Causes burns. Lung irritant. May cause delayed lung injury.

Precautions

KEEP OUT OF THE REACH OF CHILDREN. Avoid contact with eyes. Wear suitable protective eyewear. Avoid prolonged or repeated contact with skin. Wear suitable gloves. Wear suitable protective clothing. Do not breathe dust. In case of insufficient ventilation, wear suitable respiratory equipment. Wash soiled clothing before reuse.

First Aid

Wash exposed skin with soap and water. Flush eyes with large quantities of water. If breathing is difficult, move person to fresh air.

Waste Disposal Method

This product when discarded or disposed of is not listed as a hazardous waste in federal regulations. Dispose of in a landfill in accordance with local regulations.

For additional information on personal protective equipment, first aid, and emergency procedures, refer to the product Material Safety Data Sheet (MSDS) on the job site or contact the company at the address or phone numbers given below.

Proposition 65

This product contains materials listed by the state of California as known to cause cancer, birth defects, or reproductive harm.

VOC Content

0 lbs/gal or 0 g/L

For medical emergencies only, call ChemTrec (1-800-424-9300).

**BASF Construction Chemicals, LLC –
Building Systems**

889 Valley Park Drive
Shakopee, MN, 55379

www.BuildingSystems.BASF.com

Customer Service 800-433-9517
Technical Service 800-243-6739

For professional use only. Not for sale to or use by the general public.

A.3 HARRIS CONSTRUCTION GROUT MANUFACTURER'S DATA SHEET

Harris Construction Grout
DATA SHEET

50 lb. Bag
Non-shrink, Non-metallic grout

A.H. Harris & Sons, Inc.

CONSTRUCTION SUPPLIES

DESCRIPTION
Harris Construction Grout is a non-shrink, non-metallic multipurpose cement-based grout. Harris Construction Grout is formulated for a wide variety of grouting applications, from damp pack to flowable through a controlled, positive expansion.

USE
Recommended applications include grouting of pump and equipment based column base plates, anchor bolts, precast and tilt-up walls.

FEATURES / BENEFITS
- Controlled positive expansion for maximum effective bearing
- Non-metallic/non-corrosivePourable/pumpableversatility
- Excellent freeze / thaw resistance
- Can be extended with pea stone for deep applications

SPECIFICATIONS / COMPLIANCES
Harris Construction Grout meets or exceeds:
CRD C-621 "Corps of Engineers Specification for Non-Shrink Grouts"
ASTM C1107 "Standard Specification for Packaged Dry, Hydraulic-Cement Grout (Non-Shrink)" at Fluid Consistency
ASTM C827, "Standard Test Method for Change in Height at Early Ages of Cylindrical Specimens from Cementitious Mixtures"
ASTM C1090, "Standard Test Method for Measuring Changes in Height of Cylindrical Specimens from Hydraulic Cement Grouts"

APPLICATION
Preparation: Remove all dirt, oil, and loose or foreign material. Any metal in contact with grout must be free of rust, oil, grease, and other foreign matter which would limit

APPLICATION (cont.)
bond. Concrete surface must be sound and roughened to insure proper bonding. Prior to placing grout, surface should be saturated for a minimum period of four hours and preferably for twenty-four hours. Remove all excess water before placement of grout. Bolts, base plates and equipment must be secure and rigid before placement of grout.

Forms:
Allow for the continuous placement of grout. Provisions for venting to avoid air entrapment must be made. Placing from one side, provide a 45° angle in the forms toa height suitable to provide a head of grout during placement. On all sides, provide a minimum 1" (2.54 cm) horizontalclearance between the base plate and forms.Forms should be at least 1" (2.54 cm) higher than thebottom of the base plate.

Mixing:
DO NOT mix by hand. Use a mechanical mixer. For small jobs, use a 1/2" (.64 cm) low speed drill with a mortar mixing paddle. For large jobs, use a horizontal-shaft mortar mixer designed for mixing grouts.

Start with minimum water requirements. Always add wter to mixer first, then slowly add powder. Use only the amount of water required for the desired placement consistency. Mixing water requirements are noted: Stiff – 50 lbs (22.7 kg) grout mixed with 0.82 - 0.89 gal (3.1 - 3.4 liters) of water Plastic – 50 lbs (22.7 kg) grout mixed with 0.89 - 0.96 gal (3.4 - 3.6 liters) of water Flowable – 50 lbs (22.7 kg) grout mixed with 0.96 - 1.06 gal (3.6 - 4.0 liters) of water Test data and recommended water amounts are based on laboratory conditions. Actual field results may vary based on jobsite conditions.

Curing: Immediately cover with clean wet rags or burlap and keep moist until final set. After final set, remove rags and apply an

APPLICATION (cont.)

ASTM C309 curing compound, such as Harris Kurseal C309.

Deep Applications: Prewashed and graded 3/8" (1 cm)
non-reactive pea gravel must be used in applications
thicker than 6" (15.2 cm):
Up to 40% 3/8" pea gravel may be added per 55 lb (25 kg) bag of grout. Best results are obtained with approximately 25% extension.

Stiff

psi	MPa	
8,000	55.2	(at 3 days)
9,500	65.5	(at 7 days)
10,000	69.0	(at 28 days)

Plastic

psi	MPa	
6,000	41.46	(at 3 days)
7,000	48.3	(at 7 days)
8,500	58.6	(at 28 days)

Fluid

psi	MPa	
1,500	10.3	(at 1 days)
5,000	34.5	(at 3 days)
6,000	41.3	(at 7 days)
7,000	48.0	(at 28 days)

Expansion Percentage (CRD C-621)

Flowable

Expansion	
0.08%	(at 28 days)

Fluid

Expansion	
0.01%	(at 3 days)
0.01%	(at 7 days)
0.01%	(at 28 days)

LIMITATIONS

Hot Weather Conditions: Provide shade for area to be grouted. Use cool or chilled mixing water. Protect grout from direct sun exposure for up to twenty four hours after grouting. The maximum temperature (ambient and substrate) for grouting is 85°F (29°C). The maximum grout temperature should not exceed 80°F (27°F). For additional information, refer to ACI 305, Recommended Practices for Hot Weather Concreting.

Cold Weather Conditions: Raise the temperature of the area to be grouted with space heaters or steam. Warm the mixing water. Cover and insulate the

LIMITATIONS (cont.)

grout to retain warmth. The minimum temperature (ambient, substrate and grout) for grouting is 40°F (5° C). For additional information, refer to ACI 306, Recommended Practices for Cold Weather Grouting.

SHELF LIFE /STORAGE

Harris Construction Grout should be stored in a cool, dry interior area. At no time should material be exposed to high moisture, rain, or snow conditions. When stored in the original, tightly closed container, the shelf life is one year from the date of manufacture.

TECHNICAL SERVICES

For assistance, contact technical services at:
860-665-9494
www.ahharris.com
24 HOUR EMERGENCY CONTACT:
CHEMTREC - 800-424-9300

WARRANTY

NOTICE-READ CAREFULLY
CONDITIONS OF SALE
A.H.Harris offers this product for sale subject to and limited by the warranty which may only be varied by written agreement of a duly authorized corporate officer of A.H. Harris. No other representative of or for A.H. Harris is authorized to grant any warranty or to waive limitation of liability set forth below.

WARRANTY LIMITATION
A.H. Harris warrants this product to be free of manufacturing defects. If the product when purchased was defective and was within use period indicated on container or carton, when used, A.H. Harris will replace the defective product with new product
without charge to the purchaser. A.H. Harris makes no other warranty, either expressed or implied, concerning this product. There is no warranty of merchantability.
NO CLAIM OF ANY KIND SHALL BE GREATER THAN THE PURCHASE PRICE OF THE PRODUCT IN RESPECT OF WHICH DAMAGES ARECLAIMED.

INHERENT RISK
Purchaser assumes all risk associated with the use or application of the product.

A.4 EUCO CABLE GROUT PTX MANUFACTURER'S DATA SHEET

EUCO CABLE GROUT PTX
High Tolerance Cable Grout

Description
EUCO CABLE GROUT PTX is designed to produce a pumpable, non-shrink, high strength grout. It provides corrosion protection for steel cables, anchorages and rods. EUCO CABLE GROUT PTX is extremely flowable, and cured grout is similar in appearance to concrete. EUCO CABLE GROUT PTX exhibits thixotropic properties defined in PTI specifications, and can be used to repair previously grouted cables.

Primary Applications
- Pre-tensioned/post-tensioned cables and rods
- Post-tensioned ducts
- Precast wall panels
- Beams
- Columns
- Cable anchor plates

Features/Benefits
- Easy to pump or pour
- Non-shrink performance provides excellent bearing
- Flowable, high strength and self-leveling
- Aggregate free
- Pumpable for a minimum of 2 hrs @ 90°F (32°C)
- ♣ Can contribute to LEED points

Technical Information

Property	Result
Fluid Consistency	1.5 to 1.7 gal water/50 lb bag (5.7 to 6.4 L/22.7 kg)
Flow Rate (flow cone) ASTM C 939 & CRD C 621	9 to 20 seconds
Setting Time at 70°F (21°C) ASTM C 191	8 to 10 hours (will vary depending on material and ambient temperature)
Compressive Strength ASTM C 109	1 day: 2,000 psi (14 MPa) 3 days: 3,400 psi (23 MPa) 7 days: 5,500 psi (38 MPa) 28 days: 7,500 psi (52 MPa)
Hardened Height Change ASTM C 1090	24 hours: 0% to 0.1% 28 days: ≥ height at 24 hours (0.2%)
Plastic Expansion ASTM C 940	0% to 2% for up to 3 hours
Bleeding ASTM C 94 modified	0% at 5 minutes 0% at 3 hours (200 mL Gellman Filter @ 100 psi)
Chloride Permeability ASTM C 1202	28 days (30V for 6 hrs): 660 coulombs

EUCO CABLE GROUT PTX is a free flowing powder designed to be mixed with water. After mixing and placing, the color may initially appear much darker than the surrounding concrete. While this color will lighten up substantially as the grout cures, the grout may always appear somewhat darker than the surrounding concrete.

Shelf Life
2 years in original, unopened package.

Master Format #: 03 62 13

The Euclid Chemical Company
19218 Redwood Rd. • Cleveland, OH 44110
Phone: [216] 531-9222 • Toll-free: [800] 321-7628 • Fax: [216] 531-9596
www.euclidchemical.com

An RPM Company

Packaging/Yield

EUCO CABLE GROUT PTX is packaged in 50 lb (22.7 kg) bags or pails and yields 0.57 ft³ (0.016 m³) of fluid grout when mixed with 1.68 gal (6.4 L) of water.

Specifications/Compliances

- Complies with Post-Tensioning Institute Specifications (PTI)
- CRD C 621
- ASTM C 1107-05
- ASTM C 887
- ASTM C 1090

Directions for Use

If the contractor is not familiar with standard grout placement techniques, a pre-job meeting is suggested to review the project details unique to the particular job. Contact your local Euclid Chemical representative for additional information.

Mixing:

Consistency	Estimated Water Content*
Fluid	1.5 to 1.7 gal/50 lb (5.7 to 6.4 L/22.7kg)
Flowable	1.3 to 1.5 gal/50 lb (4.9 to 5.7 L/22.7kg)

* Do not add water in an amount that will cause bleeding. Do not add aggregate or cement to the grout since this action will change its precision grouting characteristics. **Note:** To minimize bleeding in vertical applications greater than twenty feet, The Euclid Chemical Company recommends a water dosage no greater than 1.54 gal/50 lb (5.8 L/22.7 kg).

Curing and Sealing: Cure all exposed grout by wet curing for 24 hours. Then, cure the grout with a high solids curing and sealing compound, such as SUPER REZ-SEAL or SUPER AQUA-CURE VOX.

Precautions/Limitations

- Clean tools and equipment with water before the material hardens.
- Do not add any admixture or fluidifiers.
- Do not use mixing water in an amount or at a temperature that will cause the mixed grout to bleed or segregate.
- Store materials in a dry place.
- Do not use material at temperatures that may cause premature freezing.
- Employ cold weather or hot weather grouting practices as the temperatures dictate.
- Rate of strength gain and setting times are significantly affected at temperature extremes.
- The Euclid Chemical Company is not responsible for stress corrosion caused by ingredients in the flushout, saturation, or mixing water, or for contaminants either in the space being grouted or from other materials used in the system.
- In all cases, consult the Material Safety Data Sheet before use.

Rev. 10.09

WARRANTY: The Euclid Chemical Company ("Euclid") solely and expressly warrants that its products shall be free from defects in materials and workmanship for one (1) year from the date of purchase. Unless authorized in writing by an officer of Euclid, no other representations or statements made by Euclid or its representatives, in writing or orally, shall alter this warranty. EUCLID MAKES NO WARRANTIES, IMPLIED OR OTHERWISE, AS TO THE MERCHANTABILITY OR FITNESS FOR ORDINARY OR PARTICULAR PURPOSES OF ITS PRODUCTS AND EXCLUDES THE SAME. If any Euclid product fails to conform with this warranty, Euclid will replace the product at no cost to Buyer. Replacement of any product shall be the sole and exclusive remedy available and buyer shall have no claim for incidental or consequential damages. Any warranty claim must be made within one (1) year from the date of the claimed breach. Euclid does not authorize anyone on its behalf to make any written or oral statements which in any way alter Euclid's installation information or instructions in its product literature or on its packaging labels. Any installation of Euclid products which fails to conform with such installation information or instructions shall void this warranty. Product demonstrations, if any, are done for illustrative purposes only and do not constitute a warranty or warranty alteration of any kind. Buyer shall be solely responsible for determining the suitability of Euclid's products for the Buyer's intended purposes.

A.5 SET 45 GROUT MANUFATURER'S DATA SHEET

The Chemical Company

PRODUCT DATA

3 03 01 00 Maintenance of Concrete

SET® 45 AND SET® 45 HW
Chemical-action repair mortar

Description
Set® 45 is a one-component magnesium phosphate-based patching and repair mortar. This concrete repair and anchoring material sets in approximately 15 minutes and takes rubber-tire traffic in 45 minutes. It comes in two formulations: Set® 45 Regular for ambient temperatures below 85° F (29° C) and Set® 45 Hot Weather for ambient temperatures ranging from 85 to 100° F (29 to 38° C).

Yield
A 50 lb (22.7 kg) bag of mixed with the required amount of water produces a volume of approximately 0.39 ft³ (0.011 m³); 60% extension using 1/2" (13 mm) rounded, sound aggregate produces approximately 0.58 ft³ (0.016 m³).

Packaging
50 lb (22.7 kg) multi-wall bags

Color
Dries to a natural gray color

Shelf Life
1 year when properly stored

Storage
Store in unopened containers in a clean, dry area between 45 and 90° F (7 and 32° C).

Features
- Single component
- Reaches 2,000 psi compressive strength in 1 hour
- Wide temperature use range
- Superior bonding
- Very low drying shrinkage
- Resistant to freeze/thaw cycles and deicing chemicals
- Only air curing required
- Thermal expansion and contraction similar to Portland cement concrete
- Sulfate resistant

Where to Use
APPLICATION
- Heavy industrial repairs
- Dowel bar replacement
- Concrete pavement joint repairs
- Full-depth structural repairs
- Setting of expansion device nosings
- Bridge deck and highway overlays
- Anchoring iron or steel bridge and balcony railings
- Commercial freezer rooms
- Truck docks
- Parking decks and ramps
- Airport runway-light installations

Benefits
Just add water and mix

Rapidly returns repairs to service

From below freezing to hot weather exposures

Bonds to concrete and masonry without a bonding agent

Improved bond to surrounding concrete

Usable in most environments

Fast, simple curing process

More permanent repairs

Stable where conventional mortars degrade

LOCATION
- Horizontal and formed vertical or overhead surfaces
- Indoor and outdoor applications

How to Apply
Surface Preparation
1. A sound substrate is essential for good repairs. Flush the area with clean water to remove all dust.

2. Any surface carbonation in the repair area will inhibit chemical bonding. Apply a pH indicator to the prepared surface to test for carbonation. If carbonation is present, abrade surface to a depth that is not carbonated.

3. Refer to International Concrete Repair Institute publication #s 03730 and 03732 for further surface preparation suggestions.

Technical Data

Composition

Set® 45 is a magnesium-phosphate patching and repair mortar.

Test Data

PROPERTY	RESULTS				TEST METHODS
Typical Compressive Strengths*, psi (MPa)					ASTM C 109, modified
	Plain Concrete 72° F (22° C)	Set® 45 Regular 72° F (22° C)	Set® 45 Regular 36° F (2° C)	Set® 45 HW 95° F (35° C)	
1 hour	—	2,000 (13.8)	—	—	
3 hour	—	5,000 (34.5)	—	3,000 (20.7)	
6 hour	—	5,000 (34.5)	1,200 (8.3)	5,000 (34.5)	
1 day	500 (3.5)	6,000 (41.4)	5,000 (34.5)	6,000 (41.4)	
3 day	1,900 (13.1)	7,000 (48.3)	7,000 (48.3)	7,000 (48.3)	
28 day	4,000 (27.6)	8,500 (58.6)	8,500 (58.6)	8,500 (55.2)	

NOTE: Only Set® 45 Regular formula, tested at 72° F (22° C), obtains 2,000 psi (13.8 MPa) compressive strength in 1 hour.

Property			Results	Test Methods
Modulus of Elasticity, psi (MPa)				ASTM C 469
		7 days	28 days	
Set® 45 Regular		4.18×10^5 (2.88×10^4)	4.55×10^5 (3.14×10^4)	
Set® 45 Hot Weather		4.90×10^5 (3.38×10^4)	5.25×10^5 (3.62×10^4)	
Freeze/thaw durability test, % RDM, 300 cycles, for Set® 45 and Set® 45 HW			80	ASTM C 666, Procedure A (modified**)
Scaling resistance to deicing chemicals, Set® 45 and Set® 45 HW				ASTM C 672
5 cycles			0	
25 cycles			0	
50 cycles			1.5 (slight scaling)	
Sulfate resistance				ASTM C 1012
Set® 45 length change after 52 weeks, %			0.09	
Type V cement mortar after 52 weeks, %			0.20	
Typical setting times, min, for Set® 45 at 72° F (22° C), and Set® 45 Hot Weather at 95° F (35° C)				Gilmore ASTM C 266, modified
Initial set			9 – 15	
Final set			10 – 20	
Coefficient of thermal expansion,*** both Set® 45 Regular and Set® 45 Hot Weather coefficients			7.15×10^{-6}/° F (12.8×10^{-6}/° C)	CRD-C 39
Flexural Strength, psi (MPa), 3 by 4 by 16" (75 by 100 by 406 mm) prisms, 1 day strength,				ASTM C 78, modified
Set® 45 mortar			550 (3.8)	
Set® 45 mortar with 3/8" (9 mm) pea gravel			600 (4.2)	
Set® 45 mortar with 3/8" (9 mm) crushed angular noncalcareous hard aggregate			650 (4.5)	

* All tests were performed with neat material (no aggregate)

**Method discontinues test when 300 cycles or an RDM of 60% is reached.

***Determined using 1 by 1 by 11" (25 mm by 25 mm by 279 mm) bars. Test was run with neat mixes (no aggregate). Extended mixes (with aggregate) produce lower coefficients of thermal expansion.

Test results are averages obtained under laboratory conditions. Expect reasonable variations.

Mixing

1. Set® 45 must be mixed, placed, and finished within 10 minutes in normal temperatures (72° F [22° C]). Only mix quantities that can be placed in 10 minutes or less.

2. Do not deviate from the following sequence; it is important for reducing mixing time and producing a consistent mix. Use a minimum 1/2" slow-speed drill and mixing paddle or an appropriately sized mortar mixer. Do not mix by hand.

3. Pour clean (potable) water into mixer. Water content is critical. Use a maximum of 4 pts (1.9 L) of water per 50 lb (22.7 kg) bag of Set® 45. Do not deviate from the recommended water content.

4. Add the powder to the water and mix for approximately 1 – 1-1/2 minutes.

5. Use neat material for patches from 1/2 – 2" (6 – 51 mm) in depth or width. For deeper patches, extend a 50 lb (22.7 kg) bag of Set® 45 HW by adding up to 30 lbs (13.6 kg) of properly graded, dust-free, hard, rounded aggregate or noncalcareous crushed angular aggregate, not exceeding 1/2" (6 mm) in accordance with ASTM C 33, #8. If aggregate is damp, reduce water content accordingly. Special procedures must be followed when angular aggregate is used. Contact your local BASF representative for more information. (Do not use calcareous aggregate made from soft limestone. Test aggregate for fizzing with 10% HCL).

Application

1. Immediately place the mixture onto the properly prepared substrate. Work the material firmly into the bottom and sides of the patch to ensure good bond.

2. Level the Set® 45 and screed to the elevation of the existing concrete. Minimal finishing is required. Match the existing concrete texture.

Curing

No curing is required, but protect from rain immediately after placing. Liquid-membrane curing compounds or plastic sheeting may be used to protect the early surface from precipitation, but never wet cure Set® 45.

For Best Performance

- Color variations are not indicators of abnormal product performance.
- Regular Set® 45 will not freeze at temperatures above -20° F (-29° C) when appropriate precautions are taken.
- Do not add sand, fine aggregate, or Portland cement to Set® 45.
- Do not use Set® 45 for patches less than 1/2" (13 mm) deep. For deep patches, use Set® 45 Hot Weather formula extended with aggregate, regardless of the temperature. Consult your BASF representative for further instructions.
- Do not use limestone aggregate.
- Water content is critical. Do not deviate from the recommended water content printed on the bag.
- Precondition these materials to approximately 70° F (21° C) for 24 hours before using.
- Protect repairs from direct sunlight, wind, and other conditions that could cause rapid drying of material.
- When mixing or placing Set® 45 in a closed area, provide adequate ventilation.
- Do not use Set® 45 as a precision machinery grout.
- Never featheredge Set® 45; for best results, always sawcut the edges of a patch.
- Prevent any moisture loss during the first 3 hours after placement. Protect Set® 45 with plastic sheeting or a curing compound in rapid-evaporation conditions.
- Do not wet cure.
- Do not place Set® 45 on a hot (90° F [32° C]), dry substrate.
- When using Set® 45 in contact with galvanized steel or aluminum, consult your local BASF sales representative.
- Make certain the most current versions of product data sheet and MSDS are being used; call Customer Service (1-800-433-9517) to verify the most current versions.
- Proper application is the responsibility of the user. Field visits by BASF personnel are for the purpose of making technical recommendations only and not for supervising or providing quality control on the jobsite.

Health and Safety

SET® 45

WARNING!

Contains silica, crystalline quartz, fly ash, magnesium oxide, phosphoric acid, monoammonium salt, iron oxide, silica, amphorous, aluminum oxide, sulfur trioxide.

Risks

Product is alkaline on contact with water and may cause injury to skin or eyes. Ingestion or inhalation of dust may cause irritation. Contains small amount of free respirable quartz which has been listed as a suspected human carcinogen by NTP and IARC. Repeated or prolonged overexposure to free respirable quartz may cause silicosis or other serious and delayed lung injury.

Precautions

Avoid contact with skin, eyes and clothing. Prevent inhalation of dust. Wash thoroughly after handling. Keep container closed when not in use. DO NOT take internally. Use only with adequate ventilation. Use impervious gloves, eye protection and if the TLV is exceeded or used in a poorly ventilated area, use NIOSH/MSHA approved respiratory protection in accordance with applicable Federal, state and local regulations.

First Aid

In case of eye contact, flush thoroughly with water for at least 15 minutes. In case of skin contact, wash affected areas with soap and water. If irritation persists, SEEK MEDICAL ATTENTION. Remove and wash contaminated clothing. If inhalation causes physical discomfort, remove to fresh air. If discomfort persists or any breathing difficulty occurs or if swallowed, SEEK IMMEDIATE MEDICAL ATTENTION.

Waste Disposal Method

This product when discarded or disposed of is not listed as a hazardous waste in federal regulations. Dispose of in a landfill in accordance with local regulations.

For additional information on personal protective equipment, first aid, and emergency procedures, refer to the product Material Safety Data Sheet (MSDS) on the job site or contact the company at the address or phone numbers given below.

Proposition 65

This product contains material listed by the State of California as known to cause cancer, birth defects or other reproductive harm.

VOC Content

0 g/L or 0 lbs/gal less water and exempt solvents.

For medical emergencies only, call ChemTrec (1-800-424-9300).

BASF Construction Chemicals, LLC – Building Systems
889 Valley Park Drive
Shakopee, MN, 55379
www.BuildingSystems.BASF.com

Customer Service 800-433-9517
Technical Service 800-243-6739

For professional use only. Not for sale to or use by the general public.

A.6 FIVE STAR HP EPOXY GROUT MANUFACTURER'S DATA SHEET

HP EPOXY GROUT
High Performance Precision Grout
Standard/High Flow

PRODUCT DESCRIPTION
Five Star HP Epoxy Grout is a high performance expansive, nonshrink, epoxy system for supporting equipment requiring precision alignment. Five Star HP Epoxy Grout is a three component, 100% solids, solvent-free system formulated to exhibit high early strength combined with the highest creep resistance at elevated temperatures. Five Star HP Epoxy Grout exhibits positive expansion when tested in accordance with ASTM C 827.

ADVANTAGES
- Permanent support for machinery requiring precision alignment
- High early strength
- Start-up in 16 hours or less
- Solvent-free clean up
- Adjustable flow for various conditions

- Expansive, nonshrink per ASTM C 827
- Superior creep resistance
- Chemically resistant
- 95% Effective Bearing Area (EBA) is typically achieved following proper grouting procedures
- Excellent adhesion to steel

USES
- High performance machinery grouting
- Crane rail grouting
- Precision alignment under dynamic load conditions
- Vibration dampening filler for rotating equipment

- Support of chemical tanks, vessels and rotating equipment
- Aggressive chemical environments
- Installation of anchors and dowels
- Wind turbine baseplates

PACKAGING AND YIELD
Five Star HP Epoxy Grout is a three component system consisting of partially filled containers of resin, hardener and polyethylene lined bags of aggregate. Five Star HP Epoxy Grout - Standard Flow includes five bags of aggregate for a unit yield of approximately 2.0 cubic feet (56.6 liters) of hardened material. When maximum flow is required, Five Star HP Epoxy Grout - High Flow is available with four bags of aggregate for a unit yield of approximately 1.75 cubic feet (49.6 liters) of hardened material.

SHELF LIFE
Two years in original unopened packaging when stored in dry conditions; high relative humidity will reduce shelf life.

TYPICAL PROPERTIES AT 70°F (21°C)				
	HP Epoxy Grout (Standard)		HP Epoxy Grout (High Flow)	
Clearances	4 to 6 inches (100 - 150 mm)		1 to 4 inches (25 - 100 mm)	
Height Change, ASTM C 827, at 90°F (32°C)	Positive Expansion		Positive Expansion	
Effective Bearing Area	95%		95%	
Creep, ASTM C 1181, 1 year 400 psi (2.8 MPa) 140°F (60°C)	1.2×10^{-3} in/in (mm/mm)		2.0×10^{-3} in/in (mm/mm)	
Tensile Strength, ASTM C 307	2400 psi (16.6 MPa)		2000 psi (13.8 MPa)	
Flexural Strength, ASTM C 580	4800 psi (33.1 MPa)		4400 psi (30.4 MPa)	
Coefficient of Expansion, ASTM C 531	17×10^{-6} in/in/°F (30×10^{-6} mm/mm/°C)		18×10^{-6} in/in/°F (32×10^{-6} mm/mm/°C)	
Bond to Concrete, ASTM C 882	Concrete Failure		Concrete Failure	
Working Time at 70°F (21°C)	60 minutes		45 minutes	
Compressive Strength, ASTM C 579 B*	Standard Compressive Strength psi (MPa)	Standard Compressive Modulus psi (MPa)	High Flow Compressive Strength psi (MPa)	High Flow Compressive Modulus psi (MPa)
16 Hours	11000 (75.9)	1.6×10^6 (11.0×10^3)	10000 (69.0)	1.5×10^6 (10.4×10^3)
1 Day	15000 (103.5)	2.0×10^6 (13.8×10^3)	14000 (96.6)	1.9×10^6 (13.1×10^3)
7 Days	16500 (113.9)	2.2×10^6 (15.2×10^3)	16000 (110.4)	2.1×10^6 (14.5×10^3)
Post cured at 140°F (60°C)	17500 (120.8)	2.5×10^6 (17.2×10^3)	17000 (117.3)	2.3×10^6 (15.9×10^3)

*Materials tested per ASTM C 579 B. Rate of loading 0.25 inches per minute. The data shown above reflects typical results based on laboratory testing under controlled conditions. Reasonable variations from the data shown above may result. Test methods are modified where applicable.

PLACEMENT GUIDELINES

1. **SURFACE PREPARATION:** All surfaces to be in contact with Five Star® HP Epoxy Grout shall be free of oil, grease, laitance and other contaminants. Concrete must be clean, sound, dry and roughened to ensure a good bond. An SSPC-SP6 commercial finish on all metal surfaces will optimize bond development to steel.
2. **FORMWORK:** Formwork shall be constructed of rigid non-absorbent materials, securely anchored, liquid-tight and strong enough to resist forces developed during grout placement. The clearance between formwork and baseplate shall be sufficient to allow for a headbox. The clearance for remaining sides shall be 1 to 2 inches (25 - 50 mm). Areas where bond is not desired must be treated with paste wax or polyethylene. Isolation joints may be necessary depending on pour dimensions. Contact the Five Star Engineering and Technical Service Center for further information.
3. **MIXING:** For optimum performance, all components should be conditioned to between 70°F and 80°F (21°C and 27°C) prior to use. Pour all Component B (hardener) into pail containing Component A (resin). Mix thoroughly by hand with a paddle or with a slow speed drill and paddle mixer to avoid air entrapment. Pour mixed liquids into mortar mixer (stationary barrel with moving blades). While mixing, slowly add Component C (aggregate) and mix only until aggregate is completely wet. Add Component C (aggregate) immediately after mixing Component A (resin) and Component B (hardener). Working time is approximately 60 minutes (45 minutes High Flow) when temperatures are at 70°F (21°C).
4. **METHODS OF PLACEMENT:** Five Star HP Epoxy Grout may be poured into place. All grout shall be placed from one side to the other, maintaining contact with the bottom of the baseplate at all times. When possible, use of a headbox is highly recommended (refer to the Five Star Technical Bulletin "Head Box and Plunger" for guidelines). For clearances greater than six inches (150 mm) or volumes more than 20 cubic feet (566 liters), use Five Star DP Epoxy Grout or call the Five Star Products Engineering and Technical Center at (800) 243-2206.
5. **POST-PLACEMENT PROCEDURES:** Final finishing should ensure material is flush with bottom edge of baseplate. Finishing of exposed surfaces is aided by using a solvent wiped trowel just before material becomes unworkable. In-service operation may begin immediately after minimum required grout strength and modulus have been achieved.
6. **CLEAN UP:** All tools and equipment may be cleaned with a water and strong detergent solution before material hardens. Sand may be used as an abrasive. A suitable solvent is required for clean up of material after hardening.

NOTE: PRIOR TO APPLICATION, READ ALL PRODUCT PACKAGING THOROUGHLY. For more detailed placement procedures, refer to Design-A-Spec™ installation guidelines or call the Five Star Products Engineering and Technical Service Center at (800) 243-2206.

CONSIDERATIONS

- Flowability and strength gain are adversely affected by lower temperatures.
- For placement temperatures below 55°F (13°C) or above 90°F (32°C), refer to Design-A-Spec™.
- To obtain bond, concrete shall be visibly free of surface moisture.
- When clearances are outside the recommended range or when exceeding maximum placement volumes, contact the Five Star Engineering and Technical Service Center.
- Do not add solvents to increase flowability.
- For continuous operating temperatures exceeding 180°F (82°C), contact the Five Star Engineering and Technical Service Center.
- Construction practices dictate concrete foundation should achieve its design strength before grouting.

CAUTION

Irritant, toxic, strong sensitizer. Contains epoxy resin and amine. This product may cause skin irritation. Do not inhale vapors. Provide adequate ventilation. Protect against contact with skin and eyes. Wear rubber gloves, long sleeve shirt, goggles with side shields. In case of contact with eyes, flush repeatedly with water and contact a physician. Areas of skin contact should be promptly washed with soap and water. Do not take internally. Keep product out of reach of children. **PRIOR TO USE, REFER TO MATERIAL SAFETY DATA SHEET.**

For worldwide availability, additional product information and technical support, contact your local Five Star distributor, local sales representative, or you may call Five Star's Engineering and Technical Service Center at (800) 243-2206.

WARRANTY: "FIVE STAR PRODUCTS INC. (FSP) PRODUCTS ARE MANUFACTURED TO BE FREE OF MANUFACTURING DEFECTS AND TO MEET FSP'S CURRENT PUBLISHED PHYSICAL PROPERTIES WHEN APPLIED IN ACCORDANCE WITH FSP'S DIRECTIONS AND TESTED IN ACCORDANCE WITH ASTM AND FSP STANDARDS. HOWEVER, SHOULD THERE BE DEFECTS OF MANUFACTURING OF ANY KIND, THE SOLE RIGHT OF THE USER WILL BE TO RETURN ALL MATERIALS ALLEGED TO BE DEFECTIVE, FREIGHT PREPAID TO FSP, FOR REPLACEMENT. THERE ARE NO OTHER WARRANTIES BY FSP OF ANY NATURE WHATSOEVER, EXPRESS OR IMPLIED, INCLUDING ANY WARRANTY OF MERCHANTABILITY OR FITNESS FOR A PARTICULAR PURPOSE IN CONNECTION WITH THIS PRODUCT. FSP SHALL NOT BE LIABLE FOR DAMAGES OF ANY SORT, INCLUDING PUNITIVE, ACTUAL, REMOTE, OR CONSEQUENTIAL DAMAGES, RESULTING FROM ANY CLAIMS OF BREACH OF CONTRACT, BREACH OF ANY WARRANTY, WHETHER EXPRESSED OR IMPLIED, INCLUDING ANY WARRANTY OF MERCHANTABILITY OR FITNESS FOR A PARTICULAR PURPOSE OR FROM ANY OTHER CAUSE WHATSOEVER. FSP SHALL ALSO NOT BE RESPONSIBLE FOR USE OF THIS PRODUCT IN A MANNER TO INFRINGE ON ANY PATENT HELD BY OTHERS."

Five Star Products, Inc.
Corporate Headquarters
750 Commerce Drive
Fairfield, CT 06825 USA
Tel: 203-336-7900 · Fax: 203-336-7930
http://www.fivestarproducts.com

©2009 Five Star Products, Inc. (07/01/09)
American Owned & Operated

A.7 LAFARGE DUCTAL JS1000 MANUFATURER'S DATA SHEET

JS1000

field-cast joint fill solutions for precast deck panel bridges

Ductal® JS1000 offers a combination of superior properties including strength, durability, fluidity and increased bond capacity. By utilizing these superior properties in conjunction with precast deck panels, engineers can create optimized solutions for advanced precast bridge deck systems – with simplified fabrication and installation processes.

Reinforced with steel fibers, Ductal® JS1000 is significantly stronger than conventional concrete and performs better in terms of abrasion and chemical resistance, freeze-thaw, carbonation and chloride ion penetration.

Because of its optimized gradation of the raw material components, Ductal® is also denser than conventional concrete. This "denseness", along with nanometer sized non-connected pores throughout its cementitious matrix, attributes to its remarkable imperviousness and durability against adverse conditions or aggressive agents.

PHYSICAL PROPERTIES

Characteristic Values for Design						
	Test Data				Design Values	
	Mean		Standard Deviation			
	MPa	psi	MPa	psi	MPa	psi
Compression	140	20,000	10	1,400	100	14,500
Flexural	30	4,300	5	700	-	-
Direct Tension f_t	8	1,160	1	145	5	725
	GPa	ksi	GPa	ksi	GPa	ksi
Youngs Modulus	50	7,200	2	300	45	6,500

JS1000

DURABILITY

Carbonation penetration depth	<0.5 mm
Freeze/thaw (after 300 cycles)	100%
Salt-scaling	<0.10 g/m^2

OTHER PROPERTIES

Density	2.4 – 2.6 S.G.
Capillary porosity (>10mm)	<1%
Total porosity	2 – 6%
Creep coefficient	0.2-0.5

COMPONENTS

A) Premix — silica fume, ground quartz, sand, cement
B) High tensile steel fibers — 0.2 mm (0.008 in) diameter x 14 mm (0.5 in) long (>2000 MPa/ 290 psi)
C) Admixture — high range water reducer/ 3rd generation
+ Water and/or ice

BATCHING

High shear mixers and an ambient temperature above 16ºC (60ºF) are recommended to successfully produce Ductal® JS1000. Onsite technical assistance by a Lafarge representative is recommended.

PLACING

Ductal® JS1000 can be placed by pouring with the use of a bucket, wheelbarrow or buggy. Any exposed Ductal® surfaces should be covered with poly or vapor barriers to prevent surface dehydration.

JOINT REINFORCING

To minimize any corrosion potential of the reinforcing between the precast panel and joints, non-corrosive rebar (such as GFRP or stainless steel) may be used. Black rebar reinforcement may also be utilized for bottom mat connection.

DESIGN

The high strength of Ductal® JS1000 allows for reduced joint widths. When designing a joint using Ductal® JS1000, the characteristic design values can be reached within 96 hours of casting -- as long as ambient temperatures above 16ºC (60ºF) are ensured. Please contact a Lafarge representative when designing joints with Ductal® JS1000.

Disclaimer: The values indicated above depend on the product characteristics, experimentation method, raw materials, formulae, manufacturing procedures and equipment used; all of which may vary. This data sheet provides no guarantee or commitment that the values set forth above will be achieved in any particular application of Ductal®. Ductal® is a registered trademark and may not be used without permission. The ultra-high performance material that is Ductal® and its various components are protected by various patents and may not be used except pursuant to the terms of a license agreement with the patent holder.

Add value with Ductal® JS1000

- improved durability
- increased usage life
- superior freeze/thaw resistance
- impact & abrasion resistance
- reduced joint width and complexity
- dimensional stability
- flexural strength
- lower permeability
- superior toughness
- low chloride ion diffusion
- ductility
- faster construction
- improved resistance to continuous flexing across the joints (from truck loads)
- reduced maintenance
- extremely low porosity
- reduced traffic disruption & risk
- cost effective

Lafarge Canada Inc.
1200, 10655 Southport Road S.W.
Calgary, Alberta, Canada T2W 4Y1
Email: ductal@lafarge-na.com • Tel: 403-271-9110
Toll free: 1-866-238-2825 • www.imagineductal.com

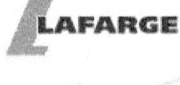

A.8 LAFARGE DUCTAL JS1100RS MANUFACTURER'S DATA SHEET

JS1100RS
Rapid Strength

JS1100RS

field-cast joint fill solutions for Accelerated Bridge Construction (ABC)

Ductal® JS1100RS offers a combination of superior properties including rapid strength, durability, fluidity and increased bond capacity. By utilizing these superior properties in conjunction with precast deck panels, engineers can create optimized solutions for advanced precast bridge deck systems – with simplified fabrication and installation processes. Compressive strengths of 55 MPa may be attained in 12 hours.

Reinforced with steel fibers, Ductal® JS1100RS is significantly stronger than conventional concrete and performs better in terms of abrasion and chemical resistance, freeze-thaw, carbonation and chloride ion penetration.

Because of its optimized gradation of the raw material components, Ductal® is also denser than conventional concrete. This "denseness", along with nanometer sized non-connected pores throughout its cementitious matrix, attributes to its remarkable imperviousness and durability against adverse conditions or aggressive agents.

PHYSICAL PROPERTIES

Characteristic Values for Design						
	Test Data				Design Values	
	Mean		Standard Deviation			
	MPa	psi	MPa	psi	MPa	psi
Compression	140	20,000	10	1,400	100	14,500
Flexural	30	4,300	5	700	-	-
Direct Tension f_t	8	1,160	1	145	5	725
	GPa	ksi	GPa	ksi	GPa	ksi
Youngs Modulus	50	7,200	2	300	45	6,500

JS1100RS

Add value with Ductal® JS1100RS

- rapid strength
- rapid set
- winter construction
- improved durability
- increased usage life
- superior freeze/thaw resistance
- impact & abrasion resistance
- reduced joint width and complexity
- dimensional stability
- flexural strength
- lower permeability
- superior toughness
- low chloride ion diffusion
- ductility
- faster construction
- improved resistance to continuous flexing across the joints (from truck loads)
- reduced maintenance
- reduced traffic disruption & risk
- cost effective

EARLY STRENGTH

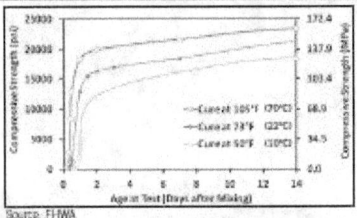

Source: FHWA

DURABILITY

• Carbonation penetration depth	<0.5 mm
• Freeze/thaw (after 300 cycles)	100%
• Salt-scaling	<0.10 g/m²

OTHER PROPERTIES

• Density	2.4 – 2.6 S.G.
• Capillary porosity (>10mm)	<1%
• Total porosity	2 – 6%
• Creep coefficient	0.2-0.5

COMPONENTS

A) Premix	- silica fume, ground quartz, sand, cement
B) High tensile steel fibers	- 0.2 mm (0.008 in) diameter x 14 mm (0.5 in) long (>2000 MPa/ 290 ksi)
C) Admixture	- high range water reducer/ 3rd generation
+ Water and/or ice	

BATCHING

High shear mixers are recommended to successfully produce Ductal® JS1100RS. Onsite technical assistance by a Lafarge representative is recommended.

PLACING

Ductal® JS1100RS can be placed by pouring with the use of a bucket, wheelbarrow or buggy. Any exposed Ductal® surfaces should be covered with poly or vapor barriers to prevent surface dehydration.

JOINT REINFORCING

To minimize any corrosion potential of the reinforcing between the precast panel and joints, non-corrosive rebar (such as GFRP or stainless steel) may be used. Black rebar reinforcement may also be utilized for bottom mat connection.

DESIGN

The high strength of Ductal® JS1100RS allows for reduced joint widths. When designing a joint using Ductal® JS1100RS, the characteristic design values can be reached within 48 hours or less depending on ambient temperatures. Please contact a Lafarge representative when designing joints with Ductal® JS1100RS.

Disclaimer: The values indicated above depend on the product characteristics, experimentation method, raw materials, formulae, manufacturing procedures and equipment used, all of which may vary. This data sheet provides no guarantee or commitment that the values set forth above will be achieved in any particular application of Ductal®. Ductal® is a registered trademark and may not be used without permission. The ultra-high performance material that is Ductal® and its various components are protected by various patents and may not be used except pursuant to the terms of a license agreement with the patent holder.

Lafarge Canada Inc.
1200, 10655 Southport Road S.W.
Calgary, Alberta, Canada T2W 4Y1
Email: ductal@lafarge-na.com • Tel: 403-271-9110
Toll free: 1-866-238-2825 • www.ductal-lafarge.com

A.9   VIRGINIA A4 MIX DESIGN PROPORTIONS

http://www.virginiadot.org/business/resources/Materials/MCS_Study_Guides/bu-mat-ConcreteCh3.pdf

TABLE II-17 Requirements for Hydraulic Cement Concrete

Class of Concrete	Design Min. Laboratory Compressive Strength at 28 days (f'c) (psi)	Aggregate Size No.[6]	Design Max. Laboratory Permeability at 28 days (Coulombs)[5]	Design Max. Laboratory Permeability at 28 days - Over tidal water (Coulombs)[5]	Nominal Max. Aggregate Size (in)	Min. Grade Aggregate	Min. Cementitious Content (lb/cu.yd)	Max. Water/ Cementitious Mat. (lb.Water/ lb. Cement)	Consistency (in of slump)	Air Content (percent)[1]
A5 Prestressed and other special designs[2]	5,000 or as specified on plans	57 or 68	1,500	1,500	1	A	635	0.40	0-4	4 1/2 ± 1 1/2
A4 General	4,000	56 or 57	2,500	2,000	1	A	635	0.45	2-4	6 1/2 ± 1 1/2
A4 Posts & rails	4,000	7, 8 or 78	2,500	2,000	0.5	A	635	0.45	2-5	7 ± 2
A3 General	3,000	56 or 57	3,500	2,000	1	A	588	0.49	1-5	6 ± 2
A3a Paving	3,000	56 or 57	3,500	3,500	1	A	564	0.49	0-3	6 ± 2
A3b Paving	3,000	357	3,500	3,500	2	A	N.A.	0.49	0-3	6 ± 2
B2 Massive or lightly reinforced	2,200	57	N.A.	N.A.	1	B	494	0.58	0-4	4 ± 2
C1 Massive Unreinforced	1,500	57	N.A.	N.A.	1	B	423	0.71	0-3	4 ± 2
T3 Tremie seal	3,000	56 or 57	N.A.	N.A.	1	A	635	0.49	3-6	4 ± 2
Latex Hydraulic cement concrete overlay[3]	3,500	7, 8 or 78	1,500	1,500	0.5	A	658	0.40	4-6	5 ± 2

www.ingramcontent.com/pod-product-compliance
Lightning Source LLC
Chambersburg PA
CBHW081826170526
45167CB00007B/2736